The Science of Discworld III: Darwin's Watch

TERRY PRATCHETT

THE CARPET PEOPLE • THE DARK SIDE OF THE SUN • STRATA
TRUCKERS • DIGGERS • WINGS • ONLY YOU CAN SAVE MANKIND
JOHNNY AND THE DEAD • JOHNNY AND THE BOMB
THE JOHNNY MAXWELL TRILOGY • JOHNNY AND THE DEAD (adapted by Stephen Briggs)
THE UNADULTERATED CAT (with Gray Jolliffe) • GOOD OMENS (with Neil Gaiman)

THE DISCWORLD ® SERIES:

THE COLOUR OF MAGIC • THE LIGHT FANTASTIC • EQUAL RITES
MORT • SOURCERY • WYRD SISTERS • PYRAMIDS • GUARDS! GUARDS!
ERIC (with Josh Kirby) • MOVING PICTURES • REAPER MAN • WITCHES ABROAD
SMALL GODS • LORDS AND LADIES • MEN AT ARMS • SOUL MUSIC
INTERESTING TIMES • MASKERADE • FEET OF CLAY • HOGFATHER • JINGO
THE LAST CONTINENT • CARPE JUGULUM • THE FIFTH ELEPHANT • THE TRUTH
THIEF OF TIME • NIGHT WATCH • MONSTROUS REGIMENT • GOING POSTAL
THE AMAZING MAURICE AND HIS EDUCATED RODENTS
THE WEE FREE MEN • A HAT FULL OF SKY
THE COLOUR OF MAGIC (graphic novel) • THE LIGHT FANTASTIC (graphic novel)
MORT: A DISCWORLD BIG COMIC (illustrated by Graham Higgins)
GUARDS! GUARDS! A DISCWORLD BIG COMIC
(adapted by Stephen Briggs, illustrated by Graham Higgins)
SOUL MUSIC: The illustrated screenplay • WYRD SISTERS: The illustrated screenplay
MORT – THE PLAY (adapted by Stephen Briggs)
WYRD SISTERS – THE PLAY (adapted by Stephen Briggs)
GUARDS! GUARDS! – THE PLAY (adapted by Stephen Briggs)
MEN AT ARMS – THE PLAY (adapted by Stephen Briggs)
MASKERADE (adapted for the stage by Stephen Briggs)
CARPE JUGULUM (adapted for the stage by Stephen Briggs)
LORDS AND LADIES (adapted for the stage by Irana Brown)
INTERESTING TIMES (adapted by Stephen Briggs)
THE FIFTH ELEPHANT (adapted by Stephen Briggs)
THE AMAZING MAURICE AND HIS EDUCATED RODENTS (adapted by Stephen Briggs)
THE STREETS OF ANKH-MORPORK (with Stephen Briggs)
THE DISCWORLD MAPP (with Stephen Briggs)
A TOURIST GUIDE TO LANCRE a Discworld Mapp (with Stephen Briggs and Paul Kidby)
DEATH'S DOMAIN (with Paul Kidby) • NANNY OGG'S COOKBOOK
THE SCIENCE OF DISCWORLD (with Ian Stewart and Jack Cohen)
THE SCIENCE OF DISCWORLD II: THE GLOBE (with Ian Stewart and Jack Cohen)
THE DISCWORD COMPANION (with Stephen Briggs)
THE PRATCHETT PORTFOLIO (with Paul Kidby) • THE LAST HERO (with Paul Kidby)

IAN STEWART

CONCEPTS OF MODERN MATHEMATICS • GAME, SET, AND MATH
THE PROBLEMS OF MATHEMATICS • DOES GOD PLAY DICE?
ANOTHER FINE MATH YOU'VE GOT ME INTO • FEARFUL SYMMETRY
NATURE'S NUMBERS • FROM HERE TO INFINITY • THE MAGICAL MAZE
LIFE'S OTHER SECRET • FLATTERLAND • WHAT SHAPE IS A SNOWFLAKE?
THE ANNOTATED FLATLAND • MATH HYSTERIA

JACK COHEN

LIVING EMBRYOS • REPRODUCTION • PARENTS MAKING PARENTS
SPERMS, ANTIBODIES AND INFERTILITY • THE PRIVILEGED APE
STOP WORKING AND START THINKING (with Graham Medley)

IAN STEWART AND JACK COHEN

THE COLLAPSE OF CHAOS • FIGMENTS OF REALITY
EVOLVING THE ALIEN (alternative title: WHAT DOES A MARTIAN LOOK LIKE?)
WHEELERS (science fiction) • HEAVEN (science fiction)

The SCIENCE of DISCWORLD III: Darwin's Watch

TERRY PRATCHETT
IAN STEWART & JACK COHEN

EBURY
PRESS

Published 2005 by Ebury Press

1 3 5 7 9 10 8 6 4 2

First published in Great Britain in 2005

EBURY PRESS
Random House, 20 Vauxhall Bridge Road, London SW1V 2SA

RANDOM HOUSE AUSTRALIA PTY LIMITED
20 Alfred Street, Milsons Point, Sydney, New South Wales 2061,
Australia

RANDOM HOUSE NEW ZEALAND LIMITED
18 Poland Road, Glenfield, Auckland 10, New Zealand

RANDOM HOUSE (PTY) LIMITED
Endulini, 5A Jubilee Road, Parktown 2193, South Africa

The Random House Group Limited Reg. No. 954009

www.randomhouse.co.uk

Papers used by Ebury Press are natural, recyclable products made
from wood grown in sustainable forests.

A CIP catalogue record for this book is available from the
British Library.

ISBN 00918 982 34

Printed and bound in Great Britain by Clays Ltd, St Ives PLC

*In crossing a heath, suppose I . . . found a watch upon the ground
. . . The inference, we think, is inevitable; that the watch must have
had a maker.*

WILLIAM PALEY
NATURAL THEOLOGY

*Divine Design, the conscious process of creation, which Paley dis-
covered, and which we now know is the explanation for the
existence and purposeful form of all life, always has purpose in
mind. If the Deity can be said to play the role of Watchmaker in
nature, He is an* all-seeing *Watchmaker.*

THE REV. CHARLES DARWIN
THEOLOGY OF SPECIES

*There is grandeur in this view of life . . . and that, whilst this planet
has gone cycling on according to the fixed law of gravity, from so
simple a beginning endless forms most beautiful and most wonderful have been, and are being, evolved.*

THE REV. RICHARD DAWKINS
THE ORIGIN OF SPECIES

*There is grandeur in this view of life . . . and that, whilst this planet
has gone cycling on according to the fixed law of gravity, from so
simple a beginning endless forms most beautiful and most wonderful have been, and are being, evolved.*

CHARLES DARWIN
THE ORIGIN OF SPECIES

*Natural selection, the blind, unconscious automatic process which
Darwin discovered, and which we now know is the explanation
for the existence and apparently purposeful form of all life, has no
purpose in mind . . . If it can be said to play the role of watchmaker
in nature, it is the* blind *watchmaker.*

RICHARD DAWKINS
THE BLIND WATCHMAKER

*In crossing a heath, suppose I found a watch upon the ground.
The inference, I think, is inevitable. Some careless chronometric
surveyor must have dropped it.*

PRESERVED J. NIGHTINGALE
WATCHES ABROAD

CONTENTS

CONCERNING
ROUNDWORLD

DISCWORLD IS REAL. It's the way worlds should work. Admittedly, it is flat and goes through space on the backs of four elephants which stand on the shell of a giant turtle, but consider the alternatives.

Consider, for example, a globular world, a mere crust upon an inferno of molten rock and iron. An accidental world, made of the wreckage of old stars, the home of life which, nevertheless, in a most unhomely fashion, is regularly scythed from its surface by ice, gas, inundation or falling rocks travelling at 20,000 miles an hour.

Such an improbable world, and the entire cosmos that surrounds it, was in fact accidentally created by the wizards of Unseen University.* It was the Dean of Unseen University who in fact destabilised the raw firmament by fiddling with it, possibly leading to the belief, if folk memory extends to sub-sub-sub-sub-atomic particle level, that it was indeed all done by somebody with a beard.

Infinite in size on the inside, but about a foot across on the outside, the universe of Roundworld is now kept in a glass globe in UU, where it has been the source of much interest and concern.

Mostly, it's the source of concern. Alarmingly, it contains no narrativium.

Narrativium is not an element in the accepted sense. It is an

---

* The greatest school of magic on the Discworld. But surely you know this?

attribute of every other element, thus turning them into, in an occult sense, molecules. Iron contains not just iron, but also the story of iron, the history of iron, the part of iron that ensures that it will continue to be iron and has an iron-like job to do and is not, for example, cheese. Without narrativium, the cosmos has no story, no purpose, no *destination*.

Nevertheless, under the ancient magical rule of As Above, So Below, the crippled universe of Roundworld strives at some level to create its own narrativium. Iron seeks out other iron. Things spin. In the absence of any gods to do the creating of life, life has managed, against the odds, to create itself. Yet the humans who have evolved on the planet believe in their hearts that there are such things as gods, magic, cosmic purpose and million-to-one chances that crop up nine times out of ten. They seek stories in the world which the world, regrettably, is not equipped to tell.

The wizards, feeling somewhat guilty about this, have intervened several times in the history of Roundworld when it seemed to them to be on the wrong track. They encouraged fish (or fish-like creatures) to leave the seas, they visited the proto-civilisations of dinosaur-descendants and crabs, they despaired at the way ice and falling comets wiped out higher life forms so often – and they found some monkeys who were obsessed with sex and were quick learners, especially if sex was involved or could, by considerable ingenuity, be made to be involved.

Again the wizards intervened, teaching them that fire was not for having sex with and in general encouraging them to get off the planet before the next big extinction.

In this they have all been guided by Hex, UU's magical thinking engine, which is immensely powerful in any case, and with Roundworld, which from Hex's point of view is a mere sub-routine of Discworld and is practically godlike, although more patient.

The wizards think they have sorted it all out. The monkeys have learned about their permanently threatened world via a type of technomancy called Science and may yet escape frozen doom.

And yet . . .

The thing about best laid plans is that they *don't* often go wrong. They *sometimes* go wrong, but not often, because of having been, as aforesaid, the best laid. The kind of plans laid by wizards, who barge in, shout a lot, try to sort it all out by lunchtime and hope for the best, on the other hand . . . well, they go wrong almost instantly.

There *is* a kind of narrativium on Roundworld, if you really look.

On Discworld, the narrativium of a fish tells it that it is a fish, was a fish, and will continue to be a fish. On Roundworld, something inside a fish tells it that it is a fish, was a fish . . . and might eventually be something else . . .

. . . perhaps.

ONE

ANY OTHER BUSINESS

 IT WAS RAINING. THIS WOULD, of course, be good for the worms.

Through the trickles that coursed down the window Charles Darwin stared at the garden.

Worms, thousands of them, out there under the soft rain, turning the detritus of winter into loam, building the soil. How . . . convenient.

The ploughs of God, he thought, and winced. It was the harrows of God that plagued him now.

Strange how the rustle of the rain sounds very much like people whispering . . .

At which point, he became aware of the beetle. It was climbing up the inside of the window, a green and blue tropical jewel.

There was another one, higher up, banging fruitlessly against the pane. One landed on his head.

The air filled up with the rattle and slither of wings. Entranced, Darwin turned to look at the glowing cloud in the corner of the room. It was forming a shape . . .

It is always useful for a university to have a Very Big Thing. It occupies the younger members, to the relief of their elders (especially if the VBT is based at some distance from the seat of learning itself) and it uses up a lot of money which would otherwise only lie around causing trouble or be spent by the sociology department or, probably, both. It also helps in pushing back boundaries, and it doesn't much matter what boundaries these are, since as any researcher will tell you it's the pushing that matters, not the boundary.

It's a good idea, too, if it's a bigger VBT than anyone else's and, in particular, since this was Unseen University, the greatest magical university in the world, if it's a bigger one than the one those bastards are building at Braseneck College.

'In fact,' said Ponder Stibbons, Head of Inadvisably Applied Magic, 'theirs is really only a QBT, or Quite Big Thing. Actually, they've had so many problems with it, it's probably only a BT!'

The senior wizards nodded happily.

'And ours is certainly bigger, is it?' said the Senior Wrangler.

'Oh, yes,' said Stibbons. 'Based on what I can determine from chatting to the people at Braseneck, ours will be capable of pushing boundaries twice as big up to three times as far.'

'I hope you haven't told them that,' said the Lecturer in Recent Runes. 'We don't want them building a . . . a . . . an EBT!'

'A what, sir?' said Ponder politely, his tone saying, 'I know about this sort of special thing and I'd rather you did not pretend that you do too.'

'Um . . . an Even Bigger Thing?' said Runes, aware that he was edging into unknown territory.

'No, sir,' said Ponder, kindly. 'The next one up would be a Great Big Thing, sir. It's been postulated that if we could ever build a GBT, we would know the mind of the Creator.'

The wizards fell silent. For a moment, a fly buzzed against the high, stone-mullioned window, with its stained-glass image of Archchancellor Sloman Discovering the Special Theory of Slood, and then, after depositing a small flyspeck on Archchancellor Sloman's nose, exited with precision though a tiny hole in one pane which had been caused two centuries ago when a stone had been thrown up by a passing cart. Originally the hole had stayed there because no one could be bothered to have it fixed, but now it stayed there because it was traditional.

The fly had been born in Unseen University and because of the high, permanent magical field, was far more intelligent that the average fly. Strangely, the field never had this effect on wizards, perhaps because most of them were more intelligent than flies in any case.

'I don't think we want to do that, do we?' said Ridcully.

'It might be considered impolite,' agreed the Chair of Indefinite Studies.

'Exactly how big would a Great Big Thing be?' said the Senior Wrangler.

'The same size as the universe, sir,' said Ponder. 'Every particle of the universe would be modelled within it, in fact.'

'Quite big, then . . .'

'Yes, sir.'

'And quite hard to find room for, I should imagine.'

'Undoubtedly, sir,' said Ponder, who had long ago given up trying to explain Big Magic to the rest of the senior faculty.

'Very well, then,' said Archchancellor Ridcully. 'Thank you for your report, Mister Stibbons.' He sniffed. 'Sounds fascinatin'. And the next item: Any Other Business.' He glared around the table. 'And since there is *no* other busi—'

'Er . . .'

This was a bad word at this point. Ridcully did not like committee business. He certainly did not like any other business.

'Well, Rincewind?' he said, glaring down the length of the table.

'Um . . .' said Rincewind. 'I think that's Professor Rincewind, sir?'

'Very well, *professor*,' said Ridcully. 'Come on, it's past time for Early Tea.'

'The world's gone wrong, Archchancellor.'

As one wizard, everyone looked out at what could be seen of the world through Archchancellor Sloman Discovering the Special Theory of Slood.

'Don't be a fool, man,' said Ridcully. 'The sun's shining! It's a nice day!'

'Not this world, sir,' said Rincewind. 'The other one.'

'What other one?' said the Archchancellor, and then his expression changed.

'Not—' he began.

'Yes, sir,' said Rincewind. 'That one. It's gone wrong. Again.'

Every organisation needs someone to do those jobs it doesn't want to do or secretly thinks don't need doing. Rincewind had nineteen of them now, including Health and Safety Officer.*

It was as Egregious Professor of Cruel and Unusual Geography that he was responsible for the Globe. These days, it was on his desk out in the gloomy cellar passage where he worked, work largely consisting of waiting until people gave him some cruel *and* unusual geography to profess.

'First question,' said Ridcully, as the faculty swept along the dank flagstones. 'Why are you working out here? What's wrong with your office?'

'It's too hot in my office, sir,' said Rincewind.

'You used to complain it was too cold!'

'Yes, sir. In the winter it is. Ice freezes on the walls, sir.'

'We give you plenty of coal, don't we?'

'Ample, sir. One bucket per day per post held, as per tradition. That's the trouble, really. I can't get the porters to understand. They won't give me *less* coal, only no coal at all. So the only way to be sure of staying warm in the winter is to keep the fire going all

* The N'tuitiv tribe of Howondaland created the post of Health and Safety Officer even before the post of Witch Doctor, and certainly before taming fire or inventing the spear. They hunt by waiting for animals to drop dead, and eat them raw.

summer, which means it's so hot in there that I can't work in – don't open the door, sir!'

Ridcully, who'd just opened the office door, slammed it again, and wiped his face with a handkerchief.

'Snug,' he said, blinking the sweat out of his eyes. Then he turned to the little globe on the desk behind him.

It was about a foot across, at least on the outside. Inside, it was infinite; most wizards have no problem with facts of this sort. It contained everything there was, for a given value of 'contained everything there was', but in its default state it focused on one tiny part of everything there was, a small planet which was, currently, covered in ice.

Ponder Stibbons swivelled the omniscope that was attached to the base of the glass dome, and stared down at the little frozen world. 'Just debris at the equator,' he reported. 'They never built the big sky-hook thing that allowed them to leave.* There must have been something we missed.'

'No, we sorted it all out,' said Ridcully. 'Remember? All the people *did* get away before the planet froze.'

'Yes, Archchancellor,' said Stibbons. 'And, then again, no.'

'If I ask you to explain that, would you tell me in words I can understand?' said Ridcully.

Ponder stared at the wall for a moment. His lips moved as he tried out sentences. 'Yes,' he said at last. 'We changed the history of the world, sending it towards a future where the people could escape before it froze. It appears that something has happened to change it back since then.'

'Again? Elves did it last time!'†

'I doubt if *they've* tried again, sir.'

'But we *know* the people left before the ice,' said the Lecturer in

---

* See *The Science of Discworld* (Ebury Press, 1999, revd 2000).
† See *The Science of Discworld II* (Ebury Press, 2002).

Recent Runes. He looked from face to face, and added uncertainly, 'Don't we?'

'We thought we knew before,' said the Dean, gloomily.

'In a way, sir,' said Ponder. 'But the Roundworld universe is somewhat . . . soft and mutable. Even though *we* can see a future happen, the past can change so that from the point of view of Roundworlders it doesn't. It's like . . . taking out the last page of a book and putting a new one in. You can still read the old page, but from the point of view of the characters, the ending has changed, or . . . possibly not.'

Ridcully slapped him on the back. 'Well done, Mr Stibbons! You didn't mention quantum even once!' he said.

'Nevertheless, I suspect it may be involved,' sighed Ponder.

TWO

PALEY'S WATCH

 THE SCENE: A RADIO CHAT-SHOW in the Bible Belt of the United States, a few years ago. The host is running a phone-in about evolution, a concept that is anathema to every God-fearing southern fundamentalist. The conversation runs something like this:

HOST: So, Jerry, what do you think about evolution? Should we take any notice of Darwin's theories?

JERRY: That Darwin guy never got a Nobel Prize, did he? If he's so great, how come he don't get no Nobel?

HOST: I think you have a very good point there, Jerry.

Such a conversation did occur, and the host was not being ironic. But Jerry's point is not quite the knock-down argument he thought it was. Charles Robert Darwin died in 1882. The first Nobel Prize was awarded in 1901.

Of course, well-meaning people are often ignorant about fine points of historical detail, and it is unfair to hold that against them. But it is perfectly fair to hold something else against them: the host and his guest didn't have their brains in gear. After all, why were they having that discussion? Because, as every God-fearing southern fundamentalist knows, virtually every scientist views Darwin as one

11

of the all-time greats. It was this assertion, in fact, that Jerry was attempting to shoot down. Now, it should be pretty obvious that winners of Nobel prizes (for science) are selected by a process that relies heavily on advice from scientists. And those, we already know, are overwhelmingly of the opinion that Darwin was somewhere near the top of the scientific tree. So if Darwin didn't get a Nobel, it couldn't have been (as listeners were intended to infer) because the committee didn't think much of his work. There had to be another reason. As it happens, the main reason was that Darwin was dead.

As this story shows, evolution is still a hot issue in the Bible Belt, where it is sometimes known as 'evilution' and generally viewed as the work of the Devil. More sophisticated religious believers – especially European ones, among them the Pope – worked out long ago that evolution poses no threat to religion: it is simply how God gets things done, in this case, the manufacture of living creatures. But the Bible-Belters, in their unsophisticated fundamentalist manner, recognise a threat, and they're right. The sophisticated reconciliation of evolution with God is a wishy-washy compromise, a cop-out. Why? Because evolution knocks an enormous hole in what otherwise might be the best argument yet devised for convincing people of the existence of God, and that is the 'argument from design'.*

The universe is awesome in its size, astonishing in its intricacy. Every part of it fits neatly with every other part. Consider an ant, an anteater, an antirrhinum. Each is perfectly suited to its role (or 'purpose'). The ant exists to be eaten by anteaters, the anteater exists to eat ants, and the antirrhinum . . . well, bees like it, and that's a good thing. Each organism shows clear evidence of 'design', as if it had been made specifically to carry out some purpose. Ants are just the right size for anteaters' tongues to lick up, anteaters have long tongues to get into ants' nests. Antirrhinums are exactly the shape to

---

* So called because it starts from the phenomenon of design and deduces the existence of a cosmic designer.

be pollinated by visiting bees. And if we observe design, then surely a designer can't be far away.

Many people find this argument compelling, especially when it is developed at length and in detail, and 'designer' is given a capital 'D'. But Darwin's 'dangerous idea', as Daniel Dennett characterised it in his book with that title, puts a very big spoke into the wheel of cosmic design. It provides an alternative, very plausible, and apparently simple process, in which there is no role for design and no need for a designer. Darwin called that process 'natural selection'; nowadays we call it 'evolution'.

There are many aspects of evolution that scientists don't yet understand. The details behind Darwin's theory are still up for grabs, and every year brings new shifts of opinion as scientists try to improve their understanding. Bible-Belters understand even less about evolution, and they typically distort it into a caricature: 'blind chance'. They have no interest whatsoever in improving their understanding. But they do understand, far better than effete Europeans, that the theory of evolution constitutes a very dangerous attack on the psychology of religious belief. Not on its substance (because anything that science discovers can be attributed to the Deity and viewed as His mechanism for bringing the associated events about) but on its attitude. Once God is removed from the day-to-day running of the planet, and installed somewhere behind DNA biochemistry and the Second Law of Thermodynamics, it is no longer so obvious that He must be fundamental to people's daily lives. In particular, there is no special reason to believe that He affects those lives in any way, or would wish to, so the fundamentalist preachers could well be out of a job. Which is how Darwin's lack of a Nobel can become a debating point on American local radio. It is also the general line along which Darwin's own thinking evolved – he began his adult life as a theology student and ended it as a somewhat tormented agnostic.

*

Seen from outside, and even more so from within, the process of scientific research is disorderly and confusing. It is tempting to deduce that scientists themselves are disorderly and confused. In a way, they are – that's what research involves. If you knew what you were doing it wouldn't be research. But that's just an apology, and there are better reasons for expecting, indeed, for valuing, that kind of confusion. The best reason is that it's an extremely effective way of understanding the world, and having a fair degree of confidence in that understanding.

In her book *Defending Science – Reason* the philosopher Susan Haack illuminates the messiness of science with a simple metaphor, the crossword puzzle. Enthusiasts know that solving a crossword puzzle is a messy business. You don't solve the clues in numerical order and write them in their proper place, converging in an orderly manner to a correct solution, unless, perhaps, it's a quick crossword and you're an expert. Instead, you attack the clues rather randomly, guided only by a vague feeling of which ones look easiest to solve (some people find anagrams easy, others hate them). You cross-check proposed answers against others, to make sure everything fits. You detect mistakes, rub them out, write in corrections.

It may not sound like a *rational* process, but the end result is entirely rational, and the checks and balances – do the answers fit the clues, do the letters all fit together? – are stringent. A few mistakes may still survive, where alternative words fit both the clue and the words that intersect them, but such errors are rare (and arguably aren't really errors, just ambiguity on the part of the compiler).

The process of scientific research, says Haack, is rather like solving a crossword puzzle. Solutions to nature's riddles arrive erratically and piecemeal. When they are cross-checked against other solutions to other riddles, sometimes the answers don't fit, and then something has to be changed. Theories that were once thought to be correct turn out to be nonsense and are thrown out. A few years ago, the best explanation of the origin of stars had one small flaw: it implied that the stars were older than the universe that contained them. At any given time, some of science's answers appear to be very solid, some less so, some are dubious . . . and some are missing entirely.

Again, it doesn't sound like a rational process, but it leads to a rational *result*. Indeed, all that cross-checking, backtracking, and revision increases our confidence in the result. Remembering, always, that nothing is proved to the hilt, nothing is *final*.

Critics often use this confused and confusing process of discovery as a reason to discredit science. Those stupid scientists can't even agree among themselves, they keep changing their minds, everything they say is provisional – why should anyone else believe such a muddle? They thereby misrepresent one of science's greatest strengths by portraying it as a weakness. A rational thinker must always be prepared to change his or her mind if the evidence requires it. In science, there is no place for dogma. Of course, many individual scientists fall short of this ideal; they are only human. Entire schools of scientific thought can get trapped in an intellectual blind alley and go into denial. On the whole, though, the errors are eventually exposed – by other scientists.

Science is not the only area of human thought to develop in this flexible way. The humanities do similar things, in their own manner. But science imposes this kind of discipline upon itself more strongly, more systematically, and more effectively, than virtually any other style of thinking. And it uses experiments as a reality check.

Religions, cults, and pseudoscientific movements do not behave like that. It is extremely rare for religious leaders to change their minds about anything that is already in their Holy Book. If your beliefs are held to be revealed truth, direct from the mouth of God, it's tricky to admit to errors. All the more credit to the Catholics, then, for admitting that in Galileo's day they got it wrong about the Earth being the centre of the universe, and until recently they got it wrong about evolution.

Religions, cults and pseudoscientific movements have a different agenda from science. Science, at its best, keeps lines of enquiry open. It is always seeking new ways to test old theories, even when they seem well established. It doesn't just look at the geology of the Grand Canyon and settle on the belief that the Earth is hundreds of millions of years old, or older. It cross-checks by taking new discoveries into account. After radioactivity was discovered, it became possible to obtain more accurate dates for geological events, and to compare those with the apparent record of sedimentation in the rocks. Many dates were then revised. When continental drift came in from left field, entirely new ways to find those dates arrived, and were quickly used. More dates were revised.

Scientists – collectively – *want* to find their mistakes, so that they can get rid of them.

Religions, cults, and pseudoscientific movements want to close down lines of enquiry. They want their followers to *stop* asking questions and accept the belief system. The difference is glaring. Suppose, for instance, that scientists became convinced that there was something worth taking seriously in the theories of Erich von Däniken, that ancient ruins and structures must have been the work of visiting aliens. They would then start asking questions. Where did the aliens come from? What sort of spaceships did they have? Why did they come here? Do ancient inscriptions suggest one kind of alien or many? What is the pattern to the visitations? Whereas believers in von Däniken's theories are satisfied with generic aliens, and ask no more. Aliens explain the ruins and structures – that's cracked it, problem solved.

Similarly, to early proponents of divine design and their modern reincarnations creationism and 'intelligent design', the latest quasi-religious fad, once we know that living creatures were created (either by God, an alien, or an unspecified intelligent designer) then the problem is solved and we need look no further. We are not encouraged to look for evidence that might disprove our beliefs. Just things that confirm them. *Accept what we tell you, don't ask questions.*

Ah, yes, but science discourages questions too, say the cults and religions. You don't take *our* views seriously, you don't allow that sort of question. You try to stop us putting our ideas into school science lessons as alternatives to your world view.

To some extent, that's true – especially the bit about science lessons. But they are *science* lessons, so they should be teaching science. Whereas the claims of the cults and the creationists, and the closet theists who espouse intelligent design, are not science. Creationism is simply a theistic belief system and offers no credible scientific evidence whatsoever for its beliefs. Evidence for alien visitations is weak, incoherent, and most of it is readily explained by entirely ordinary aspects of ancient human culture. Intelligent design *claims* evidence for its views, but those claims fall apart under even casual scientific scrutiny, as documented in the 2004 books *Why Intelligent Design Fails*, edited by Matt Young and Taner Edis, and *Debating Design*, edited by William Dembski and Michael Ruse. And when people (none of the above, we hasten to point out) claim that the Grand Canyon is evidence for Noah's flood – a notorious recent incident – it's not terribly hard to prove them wrong.

The principle of free speech implies that these views should not be suppressed, but it does not imply that they should be imported into science lessons, any more than scientific alternatives to God should be imported into the vicar's Sunday sermon. If you want to get your world view into the science lesson, you've got to establish its scientific credentials. But because cults, religions and alternative belief systems stop people asking awkward questions, there's no way they can ever get that kind of evidence. It's not only chance that is blind.

The scientific vision of the planet that is currently our only home, and of the creatures with which we share it and the universe around it, has attained its present form over thousands of years. The development of science is mostly an incremental process, a lake of understanding filled by the constant accumulation of innumerable tiny raindrops. Like the water in a lake, the pool of understanding can also evaporate again – for what we think we understand today can be exposed as nonsense tomorrow, just as what we thought we understood yesterday is exposed as nonsense today. We use the word 'understanding' rather than, 'knowledge' because science is both more than, and less than a collection of immutable facts. It is more, in that it encompasses organising principles that explain what we like to think of as facts: the strange paths of the planets in the sky make perfect sense once you understand that planets are moved by gravitational forces, and that these forces obey mathematical rules. It is less, because what may look like a fact today may turn out tomorrow to have been a misinterpretation of something else. On Discworld, where obvious things tend to be true, a tiny and insignificant Sun does indeed revolve round the grand, important world of people. We used to think our world was like that too: for centuries, it was a 'fact', and an obvious one, that the Sun revolved round the Earth.

The big organising principles of science are *theories*, coherent systems of thought that explain huge numbers of otherwise isolated facts, which have survived strenuous testing deliberately designed to break them if they do not accord with reality. They have not been merely accepted as some act of scientific faith: instead, people have tried to *falsify* them – to prove them wrong – but have so far failed. These failures do not prove that the theory is *true*, because there are always new sources of potential discord. Isaac Newton's theory of gravitation, in conjunction with his laws of motion, was – and still is – good enough to explain the movements of the planets, asteroids and other bodies of the solar system in intricate detail, with high accuracy. But in some contexts, such as black holes, it has now been replaced by Albert Einstein's theory of general relativity.

Wait a few decades, and something else will surely replace that. There are plenty of signs that all is not well at the frontiers of physics.

When cosmologists have to postulate bizarre 'dark matter' to explain why galaxies don't obey the known laws of gravity, and then throw in even weirder 'dark energy' to explain why galaxies are moving apart at an increasing rate, and when the independent evidence for these two powers of darkness is pretty much non-existent, you can *smell* the coming paradigm shift.

Most science is incremental, but some is more radical. Newton's theory was one of the great breakthroughs of science – not a shower of rain disturbing the surface of the lake, but an intellectual storm that unleashed a raging torrent. *Darwin's Watch* is about another intellectual storm: the theory of evolution. Darwin did for biology what Newton had done for physics, but in a very different way. Newton developed mathematical equations that let physicists calculate numbers and test them to many decimal places; it was a quantitative theory. Darwin's idea is expressed in words, not equations, and it describes a qualitative process, not numbers. Despite that, its influence has been at least as great as Newton's, possibly even greater. Darwin's torrent still rages today.

Evolution, then, is a theory, one of the most influential, far-reaching and important theories ever devised. In this context, it's worth pointing out that the word 'theory' is often used in a quite different sense, to mean an idea that is proposed in order to be tested. Strictly speaking, the word that should be used here is 'hypothesis', but that's such a fussy, pedantic-sounding word that people tend to avoid it. Even scientists, who should know better. 'I have a theory,' they say. No, you have a hypothesis. It will take years, possibly centuries, of stringent tests, to turn it into a theory.

The theory of evolution was once a hypothesis. Now it is a theory. Detractors seize on the word and forget its dual use. '*Only* a theory,' they say dismissively. But a true theory cannot be so easily dismissed, because it has survived so much rigorous testing. In this respect there is far more reason to take the theory of evolution seriously than any explanation of life that depends on, say, religious faith, because falsification is not high on the religious agenda. Theories, in that sense, are the best established, most credible parts of science. They are, by and large, considerably more credible than most other products of the human mind. So what these people are thinking of when they chant their dismissive slogan should actually be 'only a hypothesis'.

That was a defensible position in the early days of the theory of evolution, but today it is merely ignorant. If anything *can* be a fact, evolution is. It may have to be inferred from clues deposited in the rocks, and more recently by comparing the DNA codes of different creatures, rather than being seen directly with the naked eye in real time, but you don't need an eyewitness account to make logical deductions from evidence. The evidence, from several independent sources (such as fossils and DNA), is overwhelming. Evolution has been established so firmly that our planet makes no sense at all without it. Living creatures can, and do, change over time. The fossil record shows that they have changed substantially over long periods of time, to the extent that entirely new species have arisen. Smaller changes can be observed today, over periods as short as a year, or mere days in bacteria.

Evolution *happens.*

What remains open to dispute, especially among scientists, is *how* evolution happens. Scientific theories themselves evolve, adapting to fit new observations, new discoveries, and new interpretations of old discoveries. Theories are not carved in tablets of stone. The greatest strength of science is that when faced with sufficient evidence, scientists change their minds. Not all of them, for scientists are human and have the same failings as the rest of us, but enough of them to allow science to improve.

Even today there are diehards – not a majority, despite the noise they make, but a significant minority – who deny that evolution has ever occurred. Most of them are American, because a quirk of history (coupled with some idiosyncratic tax laws) has made evolution into a major educational issue in the United States. There, the battle between Darwin's followers and his opponents is not just about the intellectual high ground. It is about dollars and cents, and it is about who influences the hearts and minds of the next generation. The struggle masquerades as a religious and scientific one, but its essence is political. In the 1920s four American states (Arkansas, Mississippi, Oklahoma, and Tennessee) made it illegal to teach children about evolution in public schools. This law remained in place for nearly half a century: it was finally banned by the Supreme Court in 1968. This has not stopped advocates of 'creation science' from trying to find ways round that decision, or even to get it reversed. Largely, however, they have failed, and one reason is that creation 'science' is not science; it lacks intellectual rigour, it fails objective tests, and at times it is plain nutty.

It is possible to maintain that God created the Earth, and no one can prove you wrong. In that sense, it is a defensible thing to believe. Scientists may feel that this 'explanation' doesn't greatly help us understand anything, but that's their problem; for all anyone can prove, it could have happened that way. But it is not sensible to follow the Anglo-Irish prelate James Ussher's biblical chronology and maintain that the act of creation happened in 4004 BC, because there is overwhelming evidence that our planet is far older than that – 4.5 billion years rather than 6000. Either God is deliberately trying to mislead us (which is conceivable, but does not fit well with the usual religious messages, and may well be heretical) or we are standing on a very old lump of rock. Allegedly, 50 per cent of Americans believe that the Earth was created less than 10,000 years ago, which if true says something rather sad about the most expensive education system in the world.

America is fighting, all over again, a battle that was fought to a finish in Europe a century ago. The European outcome was a compromise: Pope Pius XII did accept the truth of evolution in an encyclical of 1950, but that wasn't a total victory for science.* In 1981 a successor, John Paul II, gently pointed out that 'The Bible . . . does not wish to teach how the heavens were made, but how one goes to heaven.' Science was vindicated, in that the theory of evolution was generally accepted, but religious people were free to interpret that process as God's way of making living creatures. And it's a very good way, as Darwin realised, so everyone can be happy and stop arguing. Creationists, in contrast, seem not to have appreciated that if they pin their religious beliefs to a 6000-year-old planet, they are doing themselves no favours and leaving themselves no real way out.

Darwin's Watch is about a Victorian society that never happened – well, once the wizards interfered, it stopped having happened. It is not the society that creationists are still attempting to arrange, which would be far more 'fundamentalist', full of self-righteous people telling everyone else what to do and stifling any true creativity. The real Victorian era was a paradox: a society with a very strong but rather flexible religious base, where it was taken for granted that God existed, but which gave birth to a whole series of major intellectual revolutions that led, fairly directly, to today's secular Western society. Let us not forget that even in the USA there is a constitutional separation of the state from the Church. (Strangely, the United Kingdom, which in practice is one of the most secular countries in the world – hardly anyone attends church, except for christenings, weddings, and funerals – has its own state religion, and a monarch who claims to be appointed by God. Unlike Discworld, Roundworld doesn't have to make sense.) At any rate, the real Victorians were a God-fearing race, but their society encouraged mavericks like Darwin to think outside the loop, with far-reaching consequences.

*

* According to Isaac Asimov, the most practical and dramatic victory of science over religion occurred in the seventeenth century, when churches began to put up lightning conductors.

The thread of clocks and watches runs right across the metaphorical landscape of science. Newton's vision of a solar system running according to precise mathematical 'laws' is often referred to as a 'clockwork universe'. It's not a bad image, and the orrery – a model solar system, whose cogwheels make the tiny planets revolve in some semblance of reality – does look rather like a piece of clockwork. Clocks were among the most complicated machines of the seventeenth and eighteenth centuries, and they were probably the most reliable. Even today, we say that something functions 'like clockwork'; we have yet to amend this to 'atomic accuracy'.

By the Victorian age, the epitome of reliable gadgetry had become the pocket-watch. Darwin's ideas are intimately bound up with a watch, which again plays the metaphorical role of intricate mechanical perfection. The watch in question was introduced by the clergyman William Paley, who died three years after Darwin was born. It features in the opening paragraph of Paley's great work *Natural Theology*, first published in 1802.* The best way to gain a feeling for his line of thinking is to use his own words:

> In crossing a heath, suppose I pitched my foot against a stone, and were asked how the stone came to be there; I might possibly answer, that for anything I knew to the contrary, it had lain there forever: nor would it perhaps be very easy to show the absurdity of this answer. But suppose I had found a watch upon the ground, and it should be inquired how the watch happened to be in that place; I should hardly think of the answer which I had before given, that, for anything I knew, the watch might have always been there. Yet why should not this answer serve for the watch as well as for the stone? Why is it not as admissible in the second case, as in the first? For this reason, and for no other, viz. that, when we come to inspect the watch, we

---

* It is old enough to use the elongated s's parodied in Difcworld as ∫s. We have re∫i∫ted temptation except in this footnote. Though 'manife∫tation of de∫ign' does have a bit of a cachet.

perceive (what we could not discover in the stone) that its several parts are framed and put together for a purpose, e.g. that they are so formed and adjusted as to produce motion, and that motion so regulated as to point to the hour of the day; that if the different parts had been differently shaped from what they are, of a different size from what they are, or placed after any other manner, or in any other order, than that in which they are placed, either no motion at all could have been carried on in the machine, or none which would have answered the use that is now served by it.

Paley goes on to elaborate the components of a watch, leading to the crux of his argument:

This mechanism being observed ... the inference, we think, is inevitable; that the watch must have had a maker; that there must have existed, at sometime, and at some place or other, an artificer or artificers, who formed it for the purpose which we find it actually to answer; who comprehended its construction, and designed its use.

There then follows a long series of numbered paragraphs, in which Paley qualifies his argument more carefully, extends it to cases where, for instance, some parts of the watch are missing, and dismisses several objections to his reasoning. The second chapter takes up the story by describing a hypothetical 'watch' that can produce copies of itself – a remarkable anticipation of the twentieth-century concept of a Von Neumann machine. There would still be good reason, Paley states, to infer the existence of a 'contriver'; in fact, if anything, the effect would be to enhance one's admiration for the contriver's skill. Moreover, the intelligent observer

would reflect, that though the watch before him were, in some sense, the maker of the watch which was fabricated in the course of its movements, yet it was in a very different sense from that in which a carpenter, for instance, is the maker of a chair.

He continues to develop this thought, and disposes of one possible suggestion: that, just as a stone might always have existed, for all he knew, so a watch might have always existed. That is, there might have been a chain of watches, each made by its predecessor, going back infinitely far into the past, so that there never was any first watch. However, he tells us, a watch is very different from a stone: it is contrived. Perhaps stones could always have existed: who knows? But not watches. Otherwise we would have 'contrivance, but no contriver; proofs of design, but no designer'. Rejecting this suggestion on various metaphysical grounds, Paley states:

> The conclusion which the first examination of the watch, of its works, construction, and movement, suggested, was, that it must have had, for the cause and author of that construction, an artificer, who understood its mechanism, and designed its use. This conclusion is invincible. A second examination presents us with a new discovery. The watch is found, in the course of its movement, to produce another watch, similar to itself: and not only so, but we perceive in it a system or organisation, separately calculated for that purpose. What effect would this discovery have, or ought it to have, upon our former inference? What, as hath already been said, but to increase, beyond measure, our admiration of the skill which had been employed in the formation of such a machine!

Well, we can all see where the good reverend is leading, and he homes in on his target in his third chapter. Instead of a watch, consider an eye. Not lying on a heath, but in an animal, which perhaps does lie on a heath. What he does say is: compare the eye to a telescope. There are so many similarities that we are forced to deduce that the eye was 'made for vision', just as the telescope was. Some thirty pages of anatomical description reinforce the contention that the eye must have been designed for the purpose of seeing. And the eye is just one example: consider a bird, a fish, a silkworm, or a spider. Now, finally, Paley states explicitly what all his readers knew was coming from page one:

> Were there no example in the world of contrivance except that of the eye, it would be alone sufficient to support the conclusion which we draw from it, as to the necessity of an intelligent Creator.

There we have it, in a nutshell. Living creatures are so intricate, and function so effectively, and fit together so perfectly, that they can have arisen only by design. But design implies a designer. Ergo: God exists, and it was He who created Earth's magnificent panoply of life. What more is there to say? The proof is complete.

THREE

THEOLOGY OF SPECIES

 IT WAS THREE HOURS LATER . . .

The senior wizards trod carefully in the High Energy Magic Building, partly because it wasn't their natural habitat, but also because most of the students who frequented it used the floor as a filing cabinet and, distressingly, as a larder. Pizza is quite hard to remove from a sole, especially the cheese.

In the background – always in the background in the High Energy Magic Building – was Hex, the university's thinking engine.

Occasionally, bits of it, or possibly 'him', moved. Ponder Stibbons had long ago given up trying to understand how Hex worked. Possibly Hex was the only entity in the university who understood how Hex worked.

Somewhere inside Hex magic happened. Spells were reduced, not to their component candles and wands and chants, but to *what they mean*. It happened too fast to see, and perhaps too fast to understand. All that Ponder was certain about was that life was intimately involved. When Hex was thinking deeply there was a noticeable hum from the beehives along the back wall, where slots gave them access to the outside world, and everything completely ceased to work if the ant colony was removed from its big glass maze in the heart of the machine.

Ponder had set up his magic lantern for a presentation. He liked making presentations. For a brief moment in the chaos of the universe, a presentation made everything sound as if it was organised.

'Hex has run the history of Roundworld against the last copy,' he announced, as the last wizard sat down. 'He has found significant changes beginning in what was known as the nineteenth century. Slide, please, Rincewind.' There was some muffled grumbling behind the magic lantern and a picture of a plump and elderly lady appeared on the screen. 'This lady is Queen Victoria, ruler of the Empire of the British.'

'Why is she upside down?' said the Dean.

'It could be because with a globe there is technically no right way up,' said Ponder. 'But I'm hazarding that it got put in wrong. Next slide, please. With care.' Grumble, click. 'Ah, yes, this is a steam engine. The reign of Victoria was notable for great developments in science and engineering. It was a very exciting time. Except . . . next slide, please.' Grumble, click.

'Wrong slide, that man!' said Ridcully. 'It's just blank.'

'Aha, no, sir,' said Ponder, gleefully. 'That is a dynamic way of showing you that the period I just described turns out not, in fact, to have happened. It should have, but it didn't. On *this* version of the Globe, the Empire of the British did not become as big, and the other developments were all rather muted. The great wave of discovery flattened out. The world settled down to a period of stability and peace.'

'Sounds good to me,' said Ridcully, and got a chorus of 'hear, hears' from the other wizards.

'Yes, Archchancellor,' said Ponder. 'And, then again, *no*. Getting off the planet, remember? The big freeze in five hundred years' time? No land life form surviving that was bigger than a cockroach?'

'No one bothered about that?' said Ridcully.

'Not until it was too late, sir. In that world as *we* left it, the first humans walked on the Moon less than seventy years after they flew at all.'

Ponder looked at their blank faces.

'Which was quite an achievement,' he said.

'Why? We've done *that*,' said the Dean.

Ponder sighed. 'Things are different on a globe, sir. There are no broomsticks, no magic carpets, and going to the Moon is not just a case of pushing off over the edge and trying to avoid the Turtle on the way down.'

'How did they do it, then?' said the Dean.

'Using rockets, sir.'

'The things that go up and explode with lots of coloured lights?'

'Initially, sir, but fortunately they found out how to stop them doing that. Next slide, please . . .' A picture that might have been a pair of old-fashioned pantaloons appeared on the screen. 'Ah, this is our old friend, the Trousers of Time. We all know this. It's what you get when history goes two ways. What we have to do now is find out why they split. That means I shall have to—'

'Are we near the point where you mention quantum?' said Ridcully, quickly.

'I'm afraid it is looming, sir, yes.'

Ridcully stood up, gathering his robes about him. 'Ah. I think I heard the gong for dinner, gentlemen. Just as well, really.'

The moon rose. At midnight, Ponder Stibbons read what Hex had written, wandered across the dewy lawn to the Library, woke the Librarian, and asked for a copy of a book called *The Origin of Species*.

Two hours later he went back, woke the Librarian again, and asked for *Theology of Species*. As he left with it, he heard the door being locked behind him.

Later still, he fell asleep with his face in a cold pizza and both books open on his desk, dripping with bookmarks and stray pieces of anchovy.

Beside him, Hex's writing table whirred. Twenty quill pens flashed back and forth and gyrated on spring-loaded arms, making the table look like several giant spiders on their backs. And, every minute, a page dropped onto the pile that was forming on the floor . . .

Ponder dreamed fitfully of dinosaurs trying to fly. They always splashed when they reached the bottom of the cliff.

He woke up at half past eight, read the accumulated papers, and voided a small scream.

All right, all right, he thought. There is no actual *hurry,* as such. We can change it back any time we like. That's what time travel *means*.

But although the brain can think that, the panic gland never believes it. He snatched up the books and as many notes as he could carry and hurried out.

We have heard the chimes of midnight, the saying goes. The wizards had not only heard them but also the ones at one, two and three a.m. They certainly weren't interested in hearing anything at half past eight, however. The only occupant of the tables in the Great Hall was Archchancellor Ridcully, who liked an unhealthy breakfast after his early morning run. He was alone at the trestle tables in the big hall.

'I've found it!' Ponder announced, with a certain nervous triumph, and dropped the two books in front of the astonished wizard.

'Found what?' said Ridcully. 'And mind where you're putting stuff, man! You nearly had the bacon dish over!'

'I have put my finger,' Ponder declared, 'on the precise split in the Trousers of Time!'

'Good man!' said Ridcully, reaching for the flagon of brown sauce. 'Tell me about it after breakfast, will you?'

'It's a book, sir! Two books in fact! He wrote the wrong one! Look!'

Ridcully sighed. Against the enthusiasm of wizards there was no defence. He narrowed his eyes and read the title of the book Ponder Stibbons was holding:

'*Theology of Species*. And?'

'Archchancellor, it was written by a Charles Darwin, and caused rather a row when it was published, since it purported to explain the mechanism of evolution in a manner which upset some widely held beliefs. Vested interests railed against it, but it prevailed and had a significant effect on history. Er . . . the wrong one.'

'Why? What is it about?' said Ridcully, carefully taking the top off a boiled egg.

'I've only glanced at it, Archchancellor, but it appears to describe the process of evolution as one of permanent involvement by an omnipotent deity.'

'And?' Ridcully selected a piece of toast and began to cut it into soldiers.

'That's not how it works on Roundworld, sir,' said Ponder, patiently.

'That's how it does here, more or less. There's a god who sees to it.'

'Yes, sir. But, as I am sure you will remember,' said Ponder, using the words in the sense of 'as I know you have forgotten', 'we have not found any traces of Deitium on Roundworld.'

'Well, all right,' the Archchancellor conceded. 'But I don't see why the man shouldn't have written it, even so. Good solid book, by the look of it. Took some thinkin' about, I'll be bound.'

'Yes, sir,' said Ponder. 'But the book he *should* have written . . .' he thumped another volume onto the breakfast table, '. . . was this.'

Ridcully picked it up. It had a much more colourful cover than 'Theology', and the title:

Darwin Revisited
THE ORIGIN OF SPECIES
by The Rev. Richard Dawkins

'Sir, I think I can prove that because Darwin wrote the wrong book the world took a different leg of the Trousers of Time, and humanity didn't leave the planet before the big freeze,' said Ponder, standing back.

'Why did he do that, then?' said Ridcully, mystified.

'I don't know, sir. All I know is that, until a few days ago, this Charles Darwin wrote a book that said that evolution all worked naturally, without a god. Now it turns out that he didn't. Instead, he wrote a book that said it worked *because* a god was involved at every stage.'

'And this other fella, Dawkins?'

'He said Darwin had pretty much got it right except the god part. You didn't need one, he said.'

'Didn't need a god? But it says here he's a priest of some sort!'

'Er . . . sort of, sir. In the . . . history where Charles Darwin wrote *Theology of Species*, it had become more or less compulsory to take holy orders in order to attend university. Dawkins said evolution happened all by itself.'

He shut his eyes. Ridcully alone was a much better audience than the senior faculty, who'd taken cross-purposes to the status of a fine art, but his Archchancellor was a practical, sensible man and therefore found Roundworld difficult. It wasn't a sensible place.

'You've foxed me there. *How* can it *just* happen?' said Ridcully. 'It makes no sense if there isn't *someone* who knows what's going on. There's got to be a *reason.*'

'Quite so, sir. But this is Roundworld,' said Ponder. 'Remember?'

'But surely this other feller, Dawkins, made it all right again?' Ridcully floundered. 'You did say it was the right book.'

'But at the wrong time. It was too late, sir. He didn't write *his* book until more than a hundred years later. It caused a huge row—'

'An ungodly one, I suspect?' said Ridcully cheerfully, dipping the toast in the egg.

'Haha, sir, yes. But it was still too late. Humanity was well on the road to extinction.'

Ridcully picked up *Theology* and turned it over in his hands, getting butter on it.

'Seems innocent enough,' he said. 'Gods making it all happen . . . well, that's common sense.' He held up a hand. 'I know, I know! This is Roundworld, I know. But where there's something as complicated as a watch, you know there must be a watchmaker.'

'That's what the Darwin who wrote the *Theology* book said, sir, except that he stated that the watchmaker remained part of the watch,' said Ponder.

'Oilin' it, and so forth?' said Ridcully, cheerfully.

'Sort of, sir. Metaphorically.'

'Hah!' said Ridcully. 'No wonder there was a row. Priests don't like that sort of thing. They always squirm when things get mystical.'

'Oh, the *priests*? They loved it,' said Ponder.

'What? I thought you said vested interests were against it!'

'Yes, sir. I meant the philosophers and scientists,' said Ponder Stibbons. 'The technomancers. But *they lost.*'

FOUR

PALEY
ONTOLOGY

 PALEY'S METAPHOR OF THE WATCH, alluded to by Ridcully, still remains powerful; powerful enough for Richard Dawkins to title his neo-Darwinian riposte of 1986 *The Blind Watchmaker*. Dawkins* made it clear that in his view, and those of most evolutionary biologists over the past fifty years, there was no watchmaker for living organisms, in Paley's sense: 'Paley's argument is made with passionate sincerity and is informed by the best biological scholarship of his day, but it is wrong, gloriously and utterly wrong.' But, says Dawkins, if we must give the watchmaker a role, then that role must be the process of natural selection that Darwin expounded. If so, the watchmaker has no sense of purpose: it is blind. It's a neat title but easily misunderstood, and it opens the way to replies, such as the recent book by William Dembski, *How Blind Is the Watchmaker?* Dembski is an advocate of 'intelligent design', a modern reincarnation of Paley with updated biology which repeats the old mistakes in new contexts.†

* That is to say, the Richard Dawkins of *our* leg of the famous Trousers of Time, who is, in a very definite way, not in holy orders.

† For detailed and thoughtful rebuttals of the main contentions of the intelligent design-ers, plus some responses, see Matt Young and Taner Edis, *Why Intelligent Design Fails* (Rutgers University Press, 2004), and William Dembski and Michael Ruse, *Debating Design* (Cambridge University Press, 2004). And it's only a matter of time before someone writes *How Intelligent Is the Designer?*

If you did find a watch on a heath, your first thought would probably not be that there must have been a watchmaker, but a watch-*owner*. You would either wish to get the owner's property back to them, or look guiltily around to make sure they weren't anywhere nearby before you snaffled it. Paley tells us that if we find, say, a spider on the path, then we are compelled to infer the existence of a spider-maker. But he finds no such compulsion to infer the existence of a spider-owner. Why is one human social role emphasised, but the other suppressed?

Moreover, we *know* what a watch is for, and this colours our thinking. Suppose, instead, that our nineteenth-century heath-walker chanced upon a mobile phone, left there by some careless time traveller from the future. He would probably still infer 'design' from its intricate form . . . but purpose? What conceivable purpose would a mobile phone have in the nineteenth century, with no supporting network of transmission towers? There is no way to look at a mobile phone and infer some evident purpose. If its battery has run down, it doesn't do *anything*. And if what was found on the path was a computer chip – say, the engine manager of a car – then even the element of design would be undetectable, and the chip might well be dismissed as some obscure crystalline rock. Chemical analysis would confirm the diagnosis by showing that it was mostly silicon. Of course, *we* know that these things do have a designer; but in the absence of any clear purpose, Paley's heath-walker would not be entitled to make any such inference.

In short, Paley's logic is heavily biased by what a human being would know about a watch and its maker. And his analogy breaks down when we consider other features of watches. If it doesn't even work for watches, which we do understand, there's no reason for it to apply to organisms, which we don't.

He is also rather unfair to stones.

Some of the oldest rocks in the world are found in Greenland, in a 25-mile-long band known as the Isua supracrustal belt. They are the oldest known rocks among those that have been laid down on the surface of the Earth, instead of rising from the mantle below. They are 3.8 billion years old, unless we cannot reliably make inferences from observations, in which case the evidence for cosmic design has to be thrown out along with the evidence of the rocks. We know their age because they contain tiny crystals of zircon. We mention them here because they show that Paley's lack of interest in 'stones', and his casual acceptance that they might have 'lain there forever', are unjustified. The structure of a stone is nowhere near as simple as Paley assumed. In fact, it can be just as intricate as an organism, though not as obviously 'organised'. Every stone has a story to tell.

Zircons are a case in point.

Zirconium is the 40th element in the periodic table, and zircon is zirconium sulphate. It occurs in many rocks, but usually in such tiny amounts that its presence is ignored. It is extremely hard – not as hard as diamond, but harder than the hardest steel. Jewellers sometimes use it as a diamond substitute.

Zircons, then, are found in most rocks, but in this instance the important rock is granite. Granite is an igneous rock, which wells up from the molten layers beneath the Earth's crust, forcing a path through the overlying sedimentary rock that has been deposited by wind or water. Zircons form in granite that solidifies about 12 miles (20 km) down inside the Earth. The crystals are truly tiny: one 10,000th of an inch (2 microns) is typical.

Over the last few decades we have learned that our apparently stable planet is highly dynamic, with continents that wander around over the surface, carried by gigantic 'tectonic plates' which are 60 miles (100 km) thick and float on the liquid mantle. Sometimes they even crash into each other. They move less than an inch (about 2 cm) per year, on average, and on a geological timescale that's *fast*.

The north-west of Scotland was once part of North America, when the North American plate collided with the Eurasian plate; when the plates later split apart, a piece of America was left behind, forming the Moine thrust. When plates collide, they slide over each other, often creating mountains. The highest mountains on Earth today, the Himalayas, formed when India collided with the Asian mainland. They are still rising today by more than half an inch (1.3 cm) a year, though are often weathered away faster, and India is still moving northwards.

At any rate, granite deep within the Earth may be uplifted by the collision of continental plates, to appear at the surface as part of a mountain range. Being a hard rock, it survives when the softer sedimentary rocks that surround it weather away. But eventually, even granite weathers, so the mountain erodes. The zircon crystals are even harder, so they survive weathering; they separate out from the granite, to be washed down to the coast by streams and rivers, deposited on the sandy shore, and incorporated into the next layer of sedimentary rock.

As well as being very hard, zircon is chemically very stable, and it resists most chemical changes. So, as the sediment builds up, and the zircon crystal is buried under accumulating quantities of incipient rock, the crystal is relatively immune to the increasing heat and pressure. Even when the rock is cooked by deep heat, becoming metamorphic – changing its chemical structure – the crystal of zircon survives. Its one concession to the extreme environment around it is that eventually it builds a new layer, like a skin, on its surface. This 'rim', as it is called, is roughly the same age as the surrounding rock; the inner core is far older.

Now the process may repeat. The core of zircon, with its new rim, may be pushed up with the surrounding rocks to make a new mountain range. When those mountains weather, the zircon may return to the depths, to acquire a second rim. Then a third, a fourth ... Just as tree rings indicate the growth of a tree, so 'zircon rims' reflect a sequence of mountain-building and erosion. The main difference is that each ring on a tree corresponds to a period of one year, whereas the rims on the tiny zircon crystal correspond to geological cycles that typically last hundreds of millions of years. But, just as the widths of tree rings tell us something about the climate in the years that are represented, so the zircon rims tell us something about the conditions that occurred during a given geological cycle.

By one of those neat coincidences that Paley would interpret as the Hand of God but nowadays we recognise as an inevitable consequence of the sheer richness of the universe (yes, we do see that those statements *might* be the same), the zirconium atom has the same electric charge, and is much the same size, as an atom of uranium. So uranium impurities can easily sneak into that zircon crystal. This is good for science, because uranium is radioactive. Over time, it decays into lead. If we measure the ratio of uranium to lead then we can estimate the time that has elapsed since any given part of the zircon crystal was laid down. Now we have a powerful observational tool, a geological stopwatch. And we also have a simple prediction that gives us confidence in the hypothesis that the zircon crystal forms in successive stages. Namely, the core should be the oldest part of the crystal, and successive rims should become consistently younger, in separate stages.

A typical crystal might have, say, four layers. The core might date to 3.7 billion years ago, the next to 3.6 billion years, the third to 2.6 billion years, and the last one to 2.3 billion years. So here, in a simple 'stone', we have evidence for geological cycles that last between 100 million and one billion years. The order of the ages agrees with the order in which the crystal must have been deposited. If the general scenario envisaged by geologists were wrong, then it would take only a single grain of sand to disprove it. Of course that doesn't confirm the huge geological cycles: those are deduced from other evidence. Science is a crossword puzzle.

Zircons can teach us more. It is thought that the ratio of two isotopes of carbon, carbon-12 and carbon-13, may distinguish organic sources of carbon from inorganic ones. There is carbon in the Isua formation, and the ratio there suggests that life may have existed 3.8 billion years ago, surprisingly soon after the Earth's surface solidified. But this conclusion is controversial, and many scientists are not convinced that other explanations can be excluded.

At any rate, for the Isua zircons we know that it is not an option for them to have 'lain there for ever'. Stones are far more interesting than they might seem, and anyone who knows how to read the rocks can deduce many things about their history. Paley believed that he could deduce the existence of God from the complexity of an eye. We can't get God from a zircon, but we can get vast geological cycles of mountain-building and erosion . . . and just possibly, evidence for exceedingly ancient life.

Never underestimate the humble stone. It may be a watch in disguise.

Paley's position is that what you see is what you get. The appearance is the reality. His title *Natural Theology* says as much, and his subtitle could scarcely be plainer. Organisms look designed because they *are* designed, by God; they appear to have a purpose because they *do* have a purpose: God's. Everywhere Paley looked, he saw traces of God's handiwork; everything around him was evidence for the Creator.

That kind of 'evidence' exists in such abundance that there is no difficulty in accumulating examples. Paley's central example was the eye. He noted its similarity to a telescope, and deduced that since a telescope is designed, so must an eye be. The camera did not exist in his day,* but if it had existed, he would have found even closer similarities. The eye, like a telescope or a camera, has a lens to bring

---

* Only the camera obscura, a room with a pinhole in the wall. Paley first wrote about the eye in 1802, whereas genuine photography dates from 1826.

incoming light to a sharp focus, forming an image. The eye has a retina to receive that image, just as a telescope has an observer, or a screen on to which the image is projected.

The lens of the eye is useless without the retina; the retina is useless without the lens. You can't put an eye together piecemeal – you need all of it, at once, or it can't work. Later supporters of theist explanations of life turned Paley's subtle arguments into a simplistic slogan: 'What use is half an eye?'

One reason to doubt Paley's explanation of 'design' is that in science, you very seldom get what you see. Nature is far from obvious. The waves on the ocean may seem to be travelling, but the water is mainly going round and round in tiny circles. (If it wasn't, the land would quickly be swamped.) The Sun may appear to orbit the Earth, but actually it's the other way round. Mountains, apparently solid and stable, rise and fall over geological timescales. Continents move. Stars explode. So the explanation 'it appears designed because it *is* designed' is a bit too trite, a bit too obvious, a bit too shallow. That doesn't prove it's wrong, but it gives us pause.

Darwin was one of a select group of people who realised that there might be an alternative. Instead of some cosmic designer creating the impressive organisation of organisms, that organisation might come into being of its own accord. Or, more accurately, as an inevitable consequence of the physical nature of life, and its interactions with its environment. Living creatures, Darwin suggested, are not the product of design, but of what we now call 'evolution' – a process of slow, incremental change, almost imperceptible from one generation to the next, but capable of accumulating over extensive periods of time. Evolution is a consequence of three things. One is the ability of living creatures to pass on some of their attributes to their offspring. The second is the slightly hit-and-miss nature of that ability: what they pass on is seldom a precise copy, though it usually comes close. The third is 'natural selection' – creatures that are better at survival are the ones that manage to breed, and pass on their survival attributes.

Natural selection is slow.

As an accomplished student of geology – Victorian-style field geology, where you traipse about the landscape trying to work out what rocks lie under your feet, or halfway up the next mountain, and how they got there – Darwin was well aware of the sheer abyssal depth of geological time. The record of the rocks offered compelling evidence that the Earth must be very, very old indeed: tens or hundreds of millions of years, maybe more. Today's figure of 4.5 billion years is even longer than the Victorian geologists dared imagine, but probably would not have surprised them.

Even a few million years is a very long time. Small changes can turn into huge ones over such a period of time. Imagine a species of worm four inches (10 cm) long, whose length increases by one thousandth of a per cent every year, so that even very accurate measurements would not detect any change on a yearly basis. In a hundred million years, the descendants of that worm would be 30 feet (10 m) long. From annelid to anaconda. The longest worm alive today sometimes reaches lengths of 150 feet (50m), but it is a marine worm: *Lineus longissimus*, which lives in the North Sea and can be found under boulders at low tide. Earthworms are a lot shorter, but the *Megascolecid* worms of Australia can grow to a length of 10 feet (3m), which is still impressive.

We're not suggesting that evolution happens with quite that degree of simplicity or regularity, but there's no question that geological time allows huge changes to occur by imperceptible steps. In fact, most evolutionary changes are a lot faster. Observations of 'Darwin's finches', 13 species of bird that inhabit the Galápagos Islands, reveal measurable changes from one year to the next – for example, in the average sizes of the birds' beaks.

If we want to explain the rich panoply of life on Earth, it is not enough to observe that living creatures can change as the generations pass. There must also be something that *drives* those changes in a 'creative' direction. The only driving force that Paley could

imagine was God, making conscious, intelligent choices and design-
ing them in from the beginning. Darwin was more acutely aware that
organisms can and do change from each generation to the next. Both
the fossil record and his experience with the breeding of new varieties of plants and domestic animals made that fact plain. But
breeding is also a choice imposed from outside, by the breeder, so
if anything, domestic animals look like evidence in favour of Paley.

On the other hand, no human agency ever bred dinosaurs. Does
that imply that the agency was God – or did the dinosaurs somehow
breed *themselves* into new forms? Darwin realised that there is
another kind of 'choice', imposed not by intelligent will but by cir-
cumstance and context. This is 'natural selection'. In the vast, ongoing
competition for food, living space, and the opportunity to breed,
nature will automatically favour winners over losers. Competition
introduces a kind of ratchet, which mostly moves in one direction:
towards whatever works better. So we should not be surprised that
tiny incremental changes from one generation to the next should
possess some sort of overall 'direction', or dynamic, with changes
accumulating coherently across the aeons to produce something
entirely different.

This kind of description is easily misunderstood as a kind of inbuilt
tendency towards 'progress' – ever onwards, ever upwards. Ever
more complex. Many Victorians took the message that the purpose
of evolution was to bring humanity into being. We are the highest
form of creation, we are at the top of the evolutionary tree. With us,
evolution has arrived; it will now stop, having achieved its ultimate
goal.

Rubbish. 'Works better' is not an absolute statement. It applies in
a context that is itself changing. What works better today might not
do so in a million years' time – or even tomorrow. Maybe for a time,
a bird's beak will 'work better' if it is bigger and stronger. If so, that's
how it will change. Not because the birds *know* what kind of beak
will work better: because the kind of beak that works better is the

kind that survives more effectively and is therefore more likely to be inherited by succeeding generations. But the results of the competition may change the rules of the game, so that later on, big beaks may become a disadvantage; for instance, suitable food may disappear. So now smaller beaks will win.

In short, the dynamic of evolution is not prescribed in advance: it is 'emergent'. It creates its own context, and reacts to that context, as it proceeds. So at any given time we expect to find some sensible directionality to evolutionary change, consistent over many generations, but often the universe itself only finds out what that direction is by exploring what's possible and discovering what works. Over a longer timescale, the direction itself can change. It's like a river that flows through an eroding landscape: at any given time there is a clear direction to the flow, but in the long run the passage of the river can slowly change its own course.

It is also important to appreciate that individual organisms do not compete in isolation, or against a fixed background. Billions of competitions go on all the time, and their outcome may be affected by the results of other competitions. It's not like the Olympics, where the javelin-throwers politely wait for the marathon-runners to stream past. It's more like a version of the Olympics where the javelin-throwers try to spear as many marathon-runners as they can, while the steeplechasers are trying to steal their javelins to turn each hurdle into a miniature pole vault, and the marathon-runners' main aim in life is to drink the water-jump before the steeplechasers get to it and drink it first. This is the Evolympics, where everything happens at once.

The evolutionary competitions, and their outcomes, also depend on context. Climate, in particular, plays a big role. In the Galápagos, selection for beak size in Darwin's finches depends on how many birds have what size of beak, and on what kinds of food – seeds, insects, cactus – are available and in what quantities. The amount and type of food depend on which plants and insects are competing best in the struggle to survive – not least from being eaten by finches – and breed. And all of this is played out against a background of climatic variations: wet or dry summers, wet or dry winters. Observations published in 2002 by Peter and Rosemary Grant show that the main unpredictable feature of finch evolution in the Galápagos is climate. If we could forecast the climate accurately, we could predict how the finches would evolve. But we can't predict the climate well enough, and there are reasons to think that this may never be possible.

That doesn't prevent evolution from being 'predictive', hence a science, any more than it prevents meteorology from being a science. But the evolutionary predictions are contingent upon the behaviour of the climate. They predict what will happen in what circumstances, not when it will happen.

Darwin almost certainly read Paley's masterwork as a young man, and in later life he may well have used it as a touchstone for his own, more radical and far more indirect, views. Paley succinctly expressed many of the most effective objections to Darwin's ideas, long before Darwin arrived at them. Intellectual honesty demanded that Darwin should find convincing answers to Paley. Such answers are scattered throughout Darwin's epic treatise *The Origin of Species*, though Paley's name does not appear.

In particular, Darwin found it necessary to tackle the thorny question of the eye. His answer was that although the human eye appears to be a perfected mechanism, with many interdependent parts, there are plenty of different 'eyes' in the animal kingdom, and a lot of those are relatively rudimentary. They can even be arranged in a rough progression from simple light-sensing patches to pinhole cameras to complex lenses (though this arrangement should not be interpreted as an actual evolutionary sequence). Instead of half an eye, we find an eye that is half as effective at detecting light. And this is far, far better than no eye at all.

Darwin's approach to the eye is complemented by some computer experiments published by Daniel Nilsson and Suzanne Pelger* in 1994. They studied a simple model of the evolution of a light-sensing patch of cells, whose geometry could change slightly at every 'generation', and which was equipped with the capacity to develop accessories such as a lens. In their simulations, a mere 100,000 generations were enough to transform a light-sensing patch into something approaching the human eye, including a lens whose refractive index varied from place to place, to improve its focus. The human eye possesses just such a lens. Moreover, and crucially, at every one of those 100,000 steps, the eye's ability to sense light got better.

This simulation was recently criticised on the grounds that it gets out what it puts in. It doesn't explain how those light-sensing cells can appear to begin with, or how the eye's geometry can change. And it uses a rather simplistic measure of the eye's performance. These would be important criticisms if the model were being used as some kind of proof that eyes must evolve, and as an accurate description of how they did it. However, that was never the purpose of the simulation. It had two main aims. One was to show that in the simplified context of the model, evolution constrained by natural selection could make incremental improvements and get to something resembling a real eye. It wouldn't get stuck along the way with some dead-end version of the eye that could be improved only by scrapping it and starting afresh. The second aim was to estimate the time required for such a process to take place (look at the title of the paper), on the assumption that the necessary ingredients were available.

Some of the model's assumptions are easily justified, as it happens. Light carries energy and energy affects chemical bonds, so it is not

---

* 'A pessimistic estimate of the time required for an eye to evolve', *Proceedings of the Royal Society of London* B, volume 256 (1994), pp. 53–8.

surprising that many chemicals respond to light. Evolution has an immense range of molecules to draw on – proteins specified by DNA sequences in genes. The combinatorial possibilities here are truly vast: the universe is not big enough, and has not lasted long enough, to make one molecule of each possible protein as complex as, say, haemoglobin, the oxygen-carrier in blood. It would be utterly astonishing if evolution could *not* come up with at least one light-sensing pigment, and incorporate it into a cell.

There are even some ideas of how this may have happened. In *Debating Design*, Bruce Weber and David Depew point out that light-sensitive enzyme systems can be found in bacteria, and these systems are probably very ancient. The bacteria don't use them for vision, but as part of their metabolic (energy-gaining) processes. Proteins in the human lens are very similar to metabolic enzymes found in the liver. So the proteins that make the eye did not start out as components of a system whose purpose was vision. They arose elsewhere and had quite different 'functions'. Their form and function were then selectively modified when their rudimentary light-sensing powers turned out to offer an evolutionary advantage.

Although we now know quite a lot about the genetics of the human eye, no biologist claims to know *exactly* how it evolved. The fossil record is poor, and humanoid eyes don't fossilise (though trilobite eyes do). But biologists can offer simple reasons why and how the eye could have evolved, and these alone are sufficient to demolish claims that its evolution is impossible *in principle* because the eye's components are interdependent and removing any one of them causes the eye to malfunction. The eye did not evolve one component at a time. Its structure evolved in parallel.

The instigators of more recent revivals of Paley's doctrine, albeit in less overtly theist tones, have taken on board the message of the eye as a specific case . . . but its more generic aspects seem to have eluded them. Darwin's discussion of the eye, and the Nilsson–Pelger computer experiment, are not limited to eyes. Here is the deeper message. When confronted with a complex living 'mechanism', do not assume that the only way it can evolve is component by component, piece by piece. When you see a watch, do not think of hooking up springs and adding cogwheels from some standard box of spare parts. Think more of a Salvador Dali 'soft watch' that can flow and distort, deform, split apart, and rejoin. Think of a watch whose cogwheels can change shape, grow new teeth, and whose axles and supports evolve along with the cogs so that at every stage the whole thing fits together. Think of a watch that may have started out as a paper clip, and along the way became a pogo-stick. Think not of a watch that does and always did have a single purpose, which was to tell the time. Think of a watch that once held sheets of paper together and could also be straightened out to form a tooth-pick, and which later turned out to be great for bouncing, and started to be used for measuring time only when someone noticed that its rhythmic movements could chart the passing seconds.

Yes, proponents of intelligent design understand the eye . . . but only as one example, not as the basis of a general principle. 'Oh, yes, we know all about the eye,' they say (we paraphrase). 'We're not going to ask what use half an eye is. That's simple-minded nonsense.' So instead, they ask what use half a bacterial flagellum is, and thereby repeat the identical error in a different context.

We owe this example to Michael Behe, a biochemist who was baffled by the complexity of bacterial flagella. These are the 'tails' that bacteria use to move around, tiny 'screws' like a ship's propeller, driven by a rotary molecular motor. Some forty proteins are involved in making such a motor, and if you miss any of them out, it won't work. In his 1996 *Darwin's Black Box*, Behe claimed that the only possible way to make a flagellum was to encode the whole structure, in advance, in bacterial DNA. This code could not have evolved from anything simpler, because the flagellum is 'irreducibly complex'.

An organ or biochemical system is said to be irreducibly complex if removing any of its parts causes it to fail. Behe deduced that no irreducibly complex system can evolve. The example of the bacterial flagellum quickly became a cornerstone of the intelligent design movement, and Behe's principle of irreducible complexity was promoted as an unavoidable barrier to the evolution of complex structures and functions.

There are several excellent books that debate intelligent design: we've mentioned two earlier in a footnote. It's fair to say that the antis are winning the debate hands down – even in books edited by the pros, such as *Debating Design*. Perhaps the biggest problem for the pros is that Behe's fundamental concept of 'irreducible complexity' has fatal flaws. With his definition, the deduction that an irreducibly complex system cannot evolve is valid *only* if evolution always consists of adding new parts. If that were the case, then the logic is clear. Suppose we have an irreducibly complex system, and suppose that there is an evolutionary route leading to it. Focus on the final step, where the last part is added. Then whatever came before must have been a failure, so it couldn't have existed. This is absurd: end of story.

However, evolution need not merely add identifiable components, like a factory-worker assembling a machine. It can also remove them – like a builder using scaffolding and then taking it down once it's done its job. Or the entire structure can evolve in parallel. Either possibility allows an irreducibly complex system to evolve, because the next to last step no longer has to start from a system that lacks that final, vital piece. Instead, it can start from a system with an extra piece, and remove it. Or add two vital pieces simultaneously. Nothing in Behe's definition of irreducible complexity prohibits either of these.

Moreover, 'fail' is a slippery concept: a watch that lacks hands is a failure at telling the time, but you can still use it to detonate a time-bomb, or hang it on a string to make a plumb-line. Organs and biochemical systems often change their functions as they evolve, as we've just seen in the context of the eye. No satisfactory definition of 'irreducible complexity' – one that really does constitute a barrier to evolution – has yet been suggested.

According to Kenneth Miller in *Debating Design*: 'the great irony of the flagellum's increasing acceptance as an icon of the anti-evolutionist movement is the fact that research had demolished its status as an example of irreducible complexity almost at the very moment it was first proclaimed'. Removing parts from the flagellum do *not* cause it to 'fail'. The base of the bacterial motor is remarkably similar to a system that bacteria use to attack other bacteria, the 'type III secretory system'. So here we have the basis of an entirely sensible and plausible evolutionary route to the flagellum, in which protein components *do* get added on. When you remove them again, you don't get a working flagellum – but you do get a working secretory system. The bacterial method of propulsion may well have evolved from an attack mechanism.

To their credit, proponents of intelligent design are encouraging this kind of debate, but they have not yet conceded defeat, even though their entire programme rests on shaky foundations and is collapsing in ruins. Creationists, desperate to snatch at any straw of scientific respectability for their political programme to lever religion into the American state school system,* have not yet noticed that what they are currently taking as their scientific support is falling apart at the seams. The theory of intelligent design itself is not overtly theist – indeed its proponents try very hard not to draw religious conclusions. They want the scientific arguments to be considered as science. Of course that's not going to happen, because the theist implications are a little too obvious – even to atheists.

There are some things that evolution does not explain – which will gladden the heart of anyone who feels that, Darwin notwithstanding, there are some issues that science cannot address.

* They themselves refer to this programme as the 'wedge strategy'.

It is perfectly possible to agree with Darwin and his successors that the Earth is 4.5 billion years old, and that life has evolved, by purely physical and chemical processes, from inorganic beginnings – yet still find a place for a deity. Yes, in a rich and complex universe, all these things can happen without divine intervention. But . . . how did that rich and complex universe come into being?

Here, today's cosmology offers descriptions of how (Big Bang, various recent alternatives) and when (about 13 billion years ago), but not why. String theory, a recent innovation at the frontiers of physics, makes an interesting attempt at 'why?' However, it leaves an even bigger 'why?' unanswered: *why string theory?* Science develops the consequences of physical rules ('laws'), but it doesn't explain why those rules apply, or how such a set-up came to exist.

These are deep mysteries. At the moment, and probably for ever, they are not accessible to the scientific method. Here religions come into their own, offering answers to riddles about which science chooses to remain mute.

If you want answers, they are available.

Rather a lot of different ones, in fact. Choose whichever one makes you feel most comfortable.

Feeling comfortable, however, is not a criterion recognised by science. It may make us feel warm and fuzzy, but the historical development of scientific understanding shows that, time and again, warm and fuzzy is just a polite way of saying 'wrong'.

Belief systems rely on faith, not evidence. They provide answers – but they don't provide any rational process to assess those answers. So although there are questions beyond the capacity of science to answer, that's mostly because science sets itself high standards for evidence, and holds its tongue when there isn't any. The alleged superiority of belief systems compared to science, when it comes to these deep mysteries, stems not from a failure of science, but from the willingness of belief systems to accept authority without question.

So the religious person can take comfort that his or her beliefs provide answers to deep questions of human existence that are beyond the powers of science, and the atheist can take comfort that there is absolutely no reason to expect those answers to be right. But also no way to prove them wrong, so why don't we just coexist peacefully, stay off each other's turf, and each get on with our own thing? Which is easy to say but harder to do, especially when some people refuse to stick to their own turf, and use political means, or violence, to promote their views, when rational debate long ago demolished them.

Some aspects of some belief systems are testable, of course – the Grand Canyon is not evidence for Noah's flood, unless God is having a quiet joke at our expense, which admittedly would be a very Discworld thing to do. And if He is, then all bets are off, because His revealed word in [insert your preferred Holy Book] may well be a joke too. Other aspects are not testable: the deeper issues stray into intellectual territory where, in the end, you have to settle for whatever explanation your type of mind finds convincing, or just stop asking that kind of question.

But remember: what's most interesting about your beliefs, to anyone who does not share them, is not whether you're right – it's that what you believe is a window into the workings of your mind. 'Ah, so you think like *that*, do you?'

This is where the great mystery of human existence leads, and where all explanations are true – for a given value of 'true'.

FIVE

THE WRONG TROUSERS OF TIME

THE GLASS GLOBE OF ROUNDWORLD had been installed on a pedestal in front of Hex by the time most of the senior faculty were up and milling around. They were always at a bit of a loose end when Second Breakfast had finished and it wasn't yet time for Elevenses, and this looked like entertainment.

'One asks oneself whether it really is worth saving,' said the Chair of Indefinite Studies. 'It's had huge ice ages before, hasn't it? If the humans are too stupid to leave in time, then there's bound to be another interesting species around in half a million years or so.'

'But extinction is so . . . sort of . . . *final*,' said the Lecturer in Recent Runes.

'Yes, and we created their world and helped them become intelligent,' said the Dean. 'We can't just let them freeze to death. It'd be like going on holiday and not feeding the hamster.'

A watchmaker as part of the watch, thought Ponder, adjusting the university's biggest omniscope; not just making the world, but tweaking it all the time . . .

Wizards did not believe in gods. They didn't deny their existence, of course. They just didn't believe. It was nothing personal; they weren't actually *rude* about it. Gods were a visible part of the narrativium that made things work, that gave the world its purpose. It was just that they were best avoided close up.

Roundworld had no gods that the wizards had been able to find. But one that was *built in* . . . that was a new idea. A god inside every flower and stone . . . not just a god who was everywhere, but a god who *was* everywhere.

The last chapter of *Theology of Species* had been very impressive . . .

He stood back. Hex had been busy all morning. So had the Librarian. Right now he was carefully dusting books and feeding them into Hex's hopper. Hex had mastered the secret of osmotic reading, normally only ever attempted by students.

And the Librarian had located a copy of the right *Origin of Species*, the book Darwin ought to have written. It had a picture of Darwin as a frontispiece. With a pointy hat he would have passed for a wizard anywhere. If it came to that, he could have passed for the Archchancellor.

Ponder waited until the wizards had settled down and opened their popcorn.

'Gentlemen,' he said, 'I do hope you've all read my analysis . . .?'

The wizards stared at him.

'I worked very hard on it all morning,' said Ponder. 'And it was delivered to all your offices . . . .'

There was more staring.

'It had a green cover . . . .' Ponder prompted.

The staring was quite intense now. Ponder gave in.

'Perhaps I should remind you of the important points?' he said.

The faces lit up.

'Just jog our memories,' said the Dean, cheerfully.

'I discussed alternate timelines in phase space,'* said Ponder. That was a mistake, he could see. His fellow wizards weren't stupid, but you had to be careful to shape ideas to fit the holes in their heads.

---

* Phase space, in a given context, is the space of everything that might have happened, not just what did. See *The Science of Discworld*.

'Two different legs in the Trousers of Time,' said Ponder. 'In the year 1859, by the counting system commonly in use in that part of Roundworld, a book changed the way a lot of people thought about the world. It just happened to be the wrong book—'

'Prove it,' said the Chair of Indefinite Studies.

'Pardon, sir?'

'Well, correct me if I'm wrong, but supposing *Theology of Species* was the *right* book?' said the Chair.

'It muted scientific – that is, technomantic – progress for almost a hundred years, sir,' said Ponder, wearily. 'It slowed down humanity's understanding of its place in the universe.'

'You mean that it was built by wizards and left on a shelf in a corridor?' said the Chair.

'That's only true on the outside, sir,' said Ponder. 'My point is, something happened to Mr Darwin at some time in his life that caused him to write the wrong book. And it *was* wrong. Yes, it would have been the right book here on Discworld, sir. We *know* there is a God of Evolution.'

'That's right. Skinny old chap, lives on an island,' said Ridcully. 'Decent sort, in his way. Remember? He was redesigning the elephant when we were there. With wheels, very clever. Very keen on beetles, too, as I recall.'

'So why'd Darwin write this theology book instead?' the Chair of Indefinite Studies persisted.

'I don't know, sir, but as I wrote on page 4, I'm sure you recall, it was the wrong book at exactly the right time. Nevertheless, it made sense. There was something in it for everyone. All the technomancers had to do was leave a place in their science for the local god, and all the priests had to give up were a few beliefs that none of the sensible ones believed anyway—'

'Such as what?' said the Dean.

'Well, that the world was created in a week and isn't very old,' said Ponder.

'But that's true!'

'Once again, only on the outside, Dean,' said Ponder smoothly. 'As far as we can tell, *Theology of Species* polarised intellectual opinion in a curious way. In fact, haha, it *equatorialised* it, you might say.'

'I don't think we would,' said Ridcully. 'What does the word mean?'

'Ah . . . er, on a globe, the equator is an imaginary line around the middle,' said Ponder. 'What happened was that the bulk of the tech-nomancers and the priests got behind the ideas expressed in Darwin's book, because they gave everyone pretty much what they wanted. Quite of few of the technomancers had a strong belief in the god, and most of the brighter priests could see big flaws in the dogma. Together, they were a very large and influential force. The hard-line religionists and the unbending technomancers were marginalised. Out in the cold. Polarised, in fact.' This rather neat pun, although he said it himself, failed to get even a groan of acknowledgement, so he went on: 'They didn't agree with the united group and they *certainly* didn't agree with one another. And, thus, happy compromise ruled. For well over sixty years.'

'That's nice,' said the Lecturer in Recent Runes.

'Er . . . yes, sir, and then again, no,' said Ponder. 'Technomancy doesn't work well in those circumstances. It can't make real progress by consensus. Hah, being led by a bunch of self-satisfied old men who are more interesting in big dinners than asking questions is a recipe for stagnation, anyone can see that.'

The wizards nodded sagely.

'Very true,' said Archchancellor Ridcully, narrowing his eyes. 'That was an important point which needed to be made.'

'Thank you, Archchancellor.'

'And now it needs to be apologised for.'

'Sorry, Archchancellor.'

'Good. So, Mr Stibbons, what do—'

There was a rattle from Hex's writing engine. The spidery arms wove across the paper and wrote:

+++ The Chair of Indefinite Studies is correct +++

The wizards clustered round.

'Right about what?' said Ponder.

+++ Charles Darwin of *Theology of Species* was for much of his life a Rector in the Church of England, a sub-set of the British nation +++ the computer scrawled. +++ The chief function of the priests of that religion at the time was to further the arts of archaeology, local history, lepidoptery, botany, palaeontology, geology and the making of fireworks +++

'Priests did that?' said the Dean. 'What about the praying and so on?'

+++ Some of them did that too, yes, although it was considered to be showing off. The God of the English did not require much in the way of sacrifice, only that people acted decently and kept the noise down. Being a priest in that church was a natural job for a young man of good breeding and education but no very specific talent. In the rural areas they had much free time. My calculations suggest that *Theology of Species* was the book that he was destined to write. In all the histories of third-level phase space, there is only one in which he writes *The Origin of Species* +++

'Why is that?' said Ponder.

+++ The explanation is complex +++

'Well, out with it,' said Ridcully. 'We're all sensible men here.'

Another piece of paper slid off Hex's tray. It read: +++ Yes. That is the problem. You understand that every possibility of choice gives birth to a new universe in which that choice is manifest? +++

'This is the Trousers of Time again, isn't it?' said Ridcully.

+++ Yes. Except that every leg of the Trousers of Time branches into many other legs, and so do those legs and every following leg, until everywhere is full of legs, which often pass through one another or join up again +++

'I think I'm losin' track,' said Ridcully.

+++ Yes. Language is not good at this. Even mathematics gets lost.

But a little story might work. I will tell you the story. It will be not completely inaccurate +++

'Go ahead,' said Ridcully.

+++ Imagine an unimaginably large number +++

'Right. No problem there,' said Ridcully, after the wizards had consulted among themselves.

+++ Very well +++ Hex wrote. +++ From the moment that the Roundworld universe was made, it began to split into almost identical copies of itself, billions of times a second. That unimaginably large number represents all possible Roundworld universes that there are +++

'Do all these universes really exist?' said the Dean.

+++ Impossible to prove. Assume that they do. In all those universes there are hardly any in which a man called Charles Darwin exists, takes a momentous ocean voyage, and writes a hugely influential book about the evolution of life on the planet. Nevertheless, that number is still unimaginably large +++

'But imagined by a smaller imagination?' said Ridcully. 'I mean, is it half as many as the other unimaginable number?'

+++ No. It is unimaginably large. But compared to the first number, it is unimaginably small +++

The wizards debated this in whispers.

'Very well,' said Ridcully, at last. 'Keep goin' and we'll kind of join in when we can.'

+++ Even so, it is not so unimaginable as the number of universes in which the book was *The Origin of Species*. That number is quite strange and can only be imagined at all in very unusual circumstances +++

'It's unimaginably larger?' said Ridcully.

+++ Just unimaginably unique. The number **one**. Gentlemen. All by itself. One is one and all alone. One. Yes. In third-level phase space there is only one history where he gets on the boat, completes the voyage, considers the findings and writes that book. All the other alternative Darwins either did not exist, did not stay on the boat, did not *survive* the journey, did not write any book at all or wrote, in a large number of cases, *Theology of Species* and entered the Church +++

'Boat?' said Ponder. 'What boat? What've boats got to do with it?'

+++ I explained, in the successful timeline which led to humanity leaving the planet, Mr Darwin makes a significant voyage. It is one of nineteen pivotal events in the history of the species. It is almost as important as Joshua Goddelson leaving his house by the back door in 1734 +++

'Who was he?' said Ponder. 'I don't recall the name.'

+++ A shoemaker living in Hamburg, Germany +++ wrote Hex. +++ Had he left his house by the front door that day, commercial nuclear fusion would not have been perfected 283 years later +++

'That was important, was it?' said Ridcully.

+++ Vastly. Major technomancy +++

'Did it need much in the way of shoes, then?' said Ridcully, mystified.

+++ No. But the chain of causality, though complex, is clear +++

'How hard is it to get on this boat?' said the Dean.

+++ In the case of Charles Darwin, very hard +++

'Where did it go?'

+++ It sailed from England to England. But there were crucial stops along the way. Even in those histories where he did embark on the boat, he did not complete the voyage *and* complete *The Origin of Species* in every case but one +++

'Just one version of history, you say,' said Ponder Stibbons. 'Do you know why?'

+++ Yes. It is the one where you intervene +++

'But we haven't intervened,' said Ridcully.

+++ In a primitive subjective sense this is the case. However, you are going to will have already soon +++ Hex wrote.

'What? And I am not a primitive subject, Mr Hex!'

+++ I am sorry. It is hard to convey five-dimensional ideas in a language evolved to scream defiance at the monkeys in the next tree +++

The wizards looked at one another.

'Getting a man on a ship can't be hard, surely?' said the Dean.

'Is it *dangerous* in Darwin's time?' said Rincewind.

+++ Inevitably. The centre of the Globe is an inferno, humanity is protected from being fried alive by nothing more than a skin of air and magnetic forces, and the chance of an asteroid strike is ever present +++

'I think Rincewind was referring to more immediate concerns,' said Ridcully.

+++ Understood. The major city you must visit has many squalid areas and open sewers. The river bisecting it is noxious. Your destination could be considered a high-crime drainage ditch in a dangerous and dirty world +++

'Pretty much like here, you mean?'

+++ The similarity is noticeable, yes +++

The writing arms stopped moving. Bits of Hex rattled and shook. The ants ceased their purposeful scurrying and began to mill about aimlessly in their glass tubing. Hex appeared to have something on his mind.

Then one writing arm dipped its pen into the ink and wrote, slowly:

+++ There is an additional problem. It is not clear to me why Darwin did not write *Origin* somewhere in the multiple universes without your forthcoming assistance +++

'We haven't decided that we will—' Ridcully began.

+++ But you are going to have done +++

'Well, probably—'

+++ Across the entire phase space of this world Charles Darwin did many things. He became an expert watchmaker. He ran a pottery factory. In many worlds he was a country priest. In others, he was a geologist. In yet others, he did make the important voyage and, as a result wrote *Theology of Species*. In some he began to write *The Origin of Species* only to give up. Only in one timeline was *Origin* published. This should not be possible. I detect . . . +++

+++ I detect . . . +++

The wizards waited politely.

'Yes?' said Ponder.

The single pen moved across the paper.

+++ MALIGNITY +++

SIX

BORROWED TIME

 THE EVER-BRANCHING LEGS OF the Trousers of Time are a metaphor (unless you are a quantum physicist, in which case they represent a certain mathematical view of reality) for the many paths that history might have taken if events had been slightly different. Later, we'll think about all those legs, but for now, we restrict attention to one trouser. One time-line. What exactly is time?

We know what it is on Discworld. 'Time', states *The New Discworld Companion*, 'is one of the Discworld's most secretive anthropomorphic personifications. It is hazarded that time is female (she waits for no man) but she has never been seen in the mundane worlds, having always gone somewhere else just a moment before. In her chronophonic castle, made up of endless glass rooms, she does at, er, times, materialise into a tall woman with dark hair, wearing a long red-and-black dress.'

Tick.

Even Discworld has trouble with time. In Roundworld it's worse. There was a time (there we go) when space and time were considered to be totally different things. Space had, or was, extension – it sort of spread itself around, and you could move through it at will. Within reason, maybe 20 miles (30km) a day on a good horse if the tracks weren't too muddy and the highwaymen weren't too obtrusive.

61

Tick.

Time, in contrast, moved of its own volition and took you along with it. Time just *passed*, at a fixed speed of one hour per hour, always in the direction of the future. The past had already happened, the present was happening *right now* – oops, gone already – and the future had yet to happen, but by jingo, it would, you mark my words, when it was good and ready.

Tick.

You could choose where you went in space, but you couldn't choose when you went in time. You couldn't visit the past to find out what had really happened, or visit the future to find out what fate had in store for you; you just had to wait and find out. So time was completely different from space. Space was three-dimensional, with three independent directions: left/right, back/forward, up/down. Time just *was*.

Tick.

Then along came Einstein, and time started to get mixed up with space. Time-like directions were still different from space-like ones, in some ways, but you could mix them up a bit. You could borrow time *here* and pay it back somewhere else. Even so, you couldn't head off into the future and find yourself back in your own past. That would be time travel, which played no part in physics.

Ti—

What science abhors, the arts crave. Time travel may be a physical impossibility, but it is a wonderful narrative device for writers, because it allows the story to move to past, present, or future, at will. Of course you don't need a time machine to do that – the flashback is a standard literary device. But it's fun (and respectful to narrativium) to have some kind of rationale that fits into the story itself. Victorian writers liked to use dreams; a good example is Charles Dickens's *A Christmas Carol* of 1843, with its ghosts of Christmas past, present, and yet-to-come. There is even a literary subgenre of 'timeslip romances', some of them really quite steamy. The French ones.

Time travel causes problems if you treat it as more than just a literary device. When allied to free will, it leads to paradoxes. The
ultimate cliché here is the 'grandfather paradox', which goes back to
René Barjavel's story *Le Voyageur Imprudent*. You go back in time
and kill your grandfather, but because your father is then not born,
neither are you, so you *can't* go back to kill him . . . Quite why it's
always your grandfather isn't clear (except as a sign that it's a cliché,
a low-bred form of narrativium). Killing your father or mother would
have the same paradoxical consequences. And so might the slaugh-
ter of a Cretaceous butterfly, as in Ray Bradbury's 1952 short story
'A Sound of Thunder', in which a butterfly's accidental demise at the
hands* of an unwitting time traveller changes present-day politics for
the worse.

Another celebrated time paradox is the cumulative audience paradox. Certain events, the standard one being the Crucifixion, are so
endowed with narrativium that any self-respecting time tourist will
insist on seeing them. The inevitable consequence is that anyone
who visits the Crucifixion will find Christ surrounded by thousands,
if not millions, of time travellers. A third is the perpetual investment
paradox. Put your money in a bank account in 1955, take it out in
2005, with accumulated interest, then take it back to 1955 and put it
in again . . . Be careful to use something like gold, not notes – notes
from 2005 won't be valid in 1955. Robert Silverberg's *Up the Line* is
about the Time Service, a force of time police whose job is to prevent such paradoxes from getting out of hand. A similar theme occurs
in Isaac Asimov's *The End of Eternity*.

An entire class of paradoxes arises from time loops, closed loops
of causality in which events only get started because someone comes
from the future to initiate them. For example, the easiest way for
today's humanity to get hold of a time machine is if someone is presented with one by a time traveller from the far future, when such

---

* Actually, foot.

machines have already been invented. He or she then reverse engineers the machine to find out how it works, and these principles later form the basis for the future invention of the machine. Two classic stories of this type are Robert Heinlein's 'By His Bootstraps' and 'All you Zombies', the second being noteworthy for a protagonist who becomes his own father and his own mother (via a sex change). David Gerrold took this idea to extremes in *The Man Who Folded Himself*.

Science-fiction authors are divided on whether time paradoxes always neatly unwrap themselves to produce consistent results, or whether it is genuinely possible, in their fictional setting, to change the past or the present. (No one worries much about changing the future, mind you, presumably because 'free will' amounts to precisely that. We all change the future, from what it might have been to what it actually becomes, thousands of times every day. Or so we fondly imagine.) So some authors write of attempts to kill your grandfather that, by some neat twist, bring you into existence anyway. For example, your true father was not his son at all, but a man he killed. By mistakenly eliminating the wrong grandfather, you ensure that your true father survives to sire *you*. Others, like Asimov and Silverberg, set up entire organisations dedicated to making sure that the past, hence the present, remains intact. Which may or may not work.

The paradoxes associated with time travel are part of the subject's fascination, but they do rather point towards the conclusion that time travel is a logical impossibility, let alone a physical one. So we are happy to allow the wizards of Unseen University, whose world runs on magic, the facility to wander at will up and down the Roundworld timeline, switching history from one parallel universe to another, trying to get Charles Darwin – or *somebody* – to write That Book. The wizards live in Discworld, they operate outside Roundworld constraints. But we don't really imagine that Roundworld people could do the same, without external assistance, using only Roundworld science.

Strangely, many scientists at the frontiers of today's physics don't agree. To them, time travel has become an entirely respectable* research topic, paradoxes notwithstanding. It seems that there is nothing in the 'laws' of physics, as we currently understand them, that forbids time travel. The paradoxes are apparent rather than real; they can be 'resolved' without violating physical law, as we will see in Chapter 8. That may be a flaw in today's physics, as Stephen Hawking maintains; his 'chronology protection conjecture' states that as yet unknown physical laws conspire to shut down any time machine just before it gets assembled – a built-in cosmological time cop.

On the other hand, the possibility of time travel may be a profound statement about the universe. We probably won't know for sure until we get to tackle the issue using *tomorrow's* physics. And it's worth remarking that we don't really understand *time*, let alone how to travel through it.

Although (apparently) the laws of physics do not forbid time travel, it turns out that they do make it very difficult. One theoretical scheme for achieving that goal, which involves towing black holes around very fast, requires rather more energy than is contained in the entire universe. This is a bit of a bummer, and it does seem to rule out the typical science fiction time machine, about the size of a car.†

The most extensive descriptions of Discworld time are found in *Thief of Time*. The ingredients for this novel include a member of the Guild of Clockmakers, Jeremy Clockson, who is determined to make a completely accurate clock. However, he is up against a theoretical barrier, the paradoxes of the Ephebian philosopher Xeno, which are first

* Well, let's not exaggerate. You can publish papers on it without risking losing your job. It's certainly better than publishing nothing, which definitely *will* lose you your job.
† Indeed, in the *Back to the Future* movie sequence, it was a car. A Delorean. Though it did need the assistance of a railway locomotive at one point.

mentioned in *Pyramids*. A Roundworld philosopher with an oddly similar name, Zeno of Elea, born around 490 BC, stated four paradoxes about the relation between space, time and motion. He is Xeno's Roundworld counterpart, and his paradoxes bear a curious resemblance to the Ephebian philosopher's. Xeno proved by logic alone that an arrow cannot hit a running man,* and that the tortoise is the fastest animal on the Disc.† He combined both in one experiment, by shooting an arrow at a tortoise that was racing against a hare. The arrow hit the hare by mistake, and the tortoise won, which proved that he was right. In *Pyramids*, Xeno describes the thinking behind this experiment.

''s quite simple,' said Xeno. 'Look, let's say this olive stone is an arrow and this, and this –' he cast around aimlessly – 'and this stunned seagull is the tortoise, right? Now, when you fire the arrow it goes from here to the seag— the tortoise, am I right?'

'I suppose so, but—'

'But, by this time, the seagu— the tortoise has moved on a bit, hasn't he? Am I right?'

'I suppose so,' said Teppic, helplessly. Xeno gave him a look of triumph.

'So the arrow has to go a bit further, doesn't it, to where the tortoise is now. Meanwhile the tortoise has flow— moved on, not much, I'll grant you, but it doesn't have to be much. Am I right? So the arrow has a bit further to go, but the point is that by the time it gets to where the tortoise is now the tortoise isn't there. So if the tortoise keeps moving, the arrow will never hit it. It'll keep getting closer and closer, but it'll never hit it. QED.'

---

* Provided it is fired by someone who has been in the pub since lunchtime.

† Actually this is the ambiguous puzuma, which travels at near-lightspeed (which on the Disc is about the speed of sound). If you see a puzuma, it's not there. If you hear it, it's not there either.

Zeno has a similar set-up, though he garbles it into two paradoxes. The first, called the Dichotomy, states that motion is impossible, because before you can get anywhere, you have to get halfway, and before you can get there, you have to get halfway to that, and so on for ever . . . so you have to do infinitely many things to get started, which is silly. The second, Achilles and the Tortoise, is pretty much the paradox enunciated by Xeno, but with the hare replaced by the Greek hero Achilles. Achilles runs faster than the tortoise – face it, anyone can run faster than a tortoise – but he starts a bit behind, and can never catch up because whenever he reaches the place where the tortoise was, it's moved on a bit. Like the ambiguous puzuma, by the time you get to it, it's not there. The third paradox says that a moving arrow isn't moving. Time must be divided into successive instants, and at each instant the arrow occupies a definite position, so it must be at rest. If it's always at rest, it can't move. The fourth of Zeno's paradoxes, the Moving Rows (or Stadium), is more technical to describe, but it boils down to this. Suppose three bodies are level with each other, and in the smallest instant of time one moves the smallest possible distance to the right, while the other moves the smallest possible distance to the left. Then those two bodies have moved apart by twice the smallest distance, taking the smallest instant of time to do that. So when they were just the smallest distance apart, halfway to their final destinations, time must have changed by half the smallest possible instant of time. Which would be smaller, which is crazy.

There is a serious intent to Zeno's paradoxes, and a reason why there are four of them. The Greek philosophers of Roundworld antiquity were arguing whether space and time were discrete, made up of indivisible tiny units, or continuous – infinitely divisible. Zeno's four paradoxes neatly dispose of all four combinations of continuous/discrete for space with continuous/discrete for time, neatly stuffing everyone else's theories, which is how you make your mark in philosophical circles. For instance, the Moving Rows paradox shows that having both space and time discrete is contradictory.

Zeno's paradoxes still show up today in some areas of theoretical physics and mathematics, although Achilles and the Tortoise can be dealt with by agreeing that if space and time are both continuous, then infinitely many things can (indeed must) happen in a finite time. The Arrow paradox can be resolved by noting that in the general mathematical treatment of classical mechanics, known as Hamiltonian mechanics after the great (and drunken) Irish mathematician Sir William Rowan Hamilton, the state of a body is given by two quantities, not one. As well as position it also has momentum, a disguised version of velocity. The two are related by the body's motion, but they are conceptually distinct. All you *see* is position; momentum is observable only through its effect on the subsequent positions. A body in a given position with zero momentum is not moving at that instant, and so will not *go* anywhere, whereas one in the same position with non-zero momentum – which appears identical – *is* moving, even though instantaneously it stays in the same place.

Got that?

Anyway, we were talking about *Thief of Time*, and thanks to Xeno we've not yet got past page 21. The main point is that Discworld time is malleable, so the laws of narrative imperative sometimes need a little help to make sure that the narrative does what the imperative says it should.

Tick.

Lady Myria LeJean is an Auditor of reality, who has temporarily assumed human form. Discworld is relentlessly animistic; virtually everything is conscious on some level, including basic physics. The Auditors police the laws of nature; they would very likely fine you for exceeding the speed of light. They normally take the form of small grey robes with a cowl – and nothing inside. They are the ultimate bureaucrats. LeJean points out to Jeremy that the perfect clock must be able to measure Xeno's smallest unit of time. 'It must exist, mustn't it? Consider the present. It must have a length, because one end of it is connected to the past and the other is connected to the future, and if it didn't have a length then the present couldn't exist at all. There would be no *time* for it to be the present in.'

Her views correspond rather closely to current theories of the psychology of the perception of time. Our brains perceive an 'instant' as an extended, though brief, period of time. This is analogous to the way discrete rods and cones in the retina *seem* to perceive individual points, but actually sample a small region of space. The brain accepts coarse-grained inputs and smooths them out.

LeJean is explaining Xeno to Jeremy because she has a hidden agenda: if Jeremy succeeds in making the perfect clock, then time will stop. This will make the Auditors' task as clerks of the universe much simpler, because humans are always moving things around, which makes it difficult to keep track of their locations in time and space.

Tick.

Near the Discworld Hub, in a high, green valley, lies the monastery of Oi Dong, where live the fighting monks of the order of Wen, otherwise known as History Monks. They have taken upon themselves the task of ensuring that the right history happens in the right order. The monks know what is right because they guard the History Books, which are not records of what did happen, but instructions for what should.

A youngster named Ludd, a foundling brought up by the Thieves' Guild, where he was an exceptionally talented student, has been recruited to the ranks of the History Monks and given the name Lobsang. The monks' main technological aids are procrastinators, huge spinning machines that store and move time. With a procrastinator, you can borrow time and pay it back later. Lobsang wouldn't *dream* of living on borrowed time, though – but if it wasn't nailed down, he would almost certainly steal it. He can steal anything, and usually does. And, thanks to the procrastinators, time is *not* nailed down.

If you haven't got the joke by now, take another look at the title.

LeJean's plan works; Jeremy builds his clock.

Ti—

Time stops, which is what the Auditors wanted. Not only on Discworld: temporal stasis expands across the universe at the speed of light. Soon, *everything* will stop. The History Monks are powerless, for they, too, have stopped. Only Susan Sto Helit, Death's granddaughter, can get time started again. And Ronnie Soak, who used to be Kaos, the Fifth Horseman of the Apocralypse, but left because of artistic disputes before they became famous ... Fortunately, the Auditors like obeying rules, and DO NOT FEED THE ELEPHANT really perplexes them when there is no elephant to feed. Fatally, they also have a love–hate relationship with chocolate. They are living on stolen time.

A procrastinator is a sort of time machine, but it moves time itself, instead of moving people through time. Moreover, it's fact, not fiction, as is all of Discworld to those who live there. On Roundworld, the first fictional time machine, as opposed to dreams or narrative timeslip, seems to have been invented by Edward Mitchell, an editor for the *New York Sun* newspaper. In 1881 he published an anonymous story, 'The Clock That Went Backward', in his paper. The most celebrated time-travel gadget appears in Herbert George Wells's novel *The Time Machine* of 1895, and this set a standard for all that followed. The novel tells of a Victorian inventor who builds a time machine and travels into the far future. There he finds that humanity has speciated into two distinct types – the nasty Morlocks, who live deep inside caverns, and the ethereal Eloi, who are preyed on by the Morlocks and are too indolent to do anything about it. Several movies, all fairly ghastly, have been based on the book.

The novel had inauspicious beginnings. Wells studied biology, mathematics, physics, geology, drawing, and astrophysics at the Normal School of Science, which became the Royal College of Science and eventually merged with Imperial College of Science and

Technology. While a student there, he began the work that led up to *The Time Machine*. His first time-travel story 'The Chronic Argonauts' appeared in 1888 in the *Science Schools Journal*, which Wells helped to found. The protagonist voyages into the past and commits a murder. The story offers no rationale for time travel and is more of a mad-scientist tale in the tradition of Mary Shelley's *Frankenstein*, but nowhere near as well written. Wells later destroyed every copy of it he could locate, because it embarrassed him so much. It lacked even the paradoxical element of the 1891 *Tourmalin's Time Cheques* by Thomas Anstey Guthrie, which introduced many of the standard time-travel paradoxes.

Over the following three years, Wells produced two more versions of his time-travel story, now lost, but along the way the storyline mutated into a far-future vision of the human race. The next version appeared in 1894 in the *National Observer* magazine, as three connected tales with the title 'The Time Machine'. This version has many features in common with the final novel, but before publication was complete, the editor of the magazine moved to the *New Review*. There he commissioned the same series again, but this time Wells made substantial changes. The manuscripts include many scenes that were never printed: the hero journeys into the past, running into a prehistoric hippopotamus* and meeting the Puritans in 1645. The published magazine version is very similar to the one that appeared in book form in 1895. In this version the Time Traveller moves only into the future, where he finds out what will happen to the human race, which splits into the languid Eloi and the horrid Morlocks – both equally distasteful.

---

* As one does. Palaeontologists have just announced that they have found remarkably well-preserved fossils in an East Anglian quarry, showing that giant hippos weighing six or seven tons – roughly twice the weight of modern hippos – wallowed in the rivers of Norfolk 600,000 years ago. It was a warm period sandwiched between two ice ages, probably a few degrees warmer than the present day (you can tell that from insect fossils) and hyenas prowled the banks in search of carrion.

Where did Wells get the idea? The standard SF writer's reply to this question is that 'you make it up', but we have some fairly specific information in this case. In a foreword to the 1932 edition, Wells says that he was motivated by 'student discussions in the laboratories and debating society of the Royal College of Science in the eighties'. According to Wells's son, the idea came from a paper on the fourth dimension read by another student. In the introduction to the novel, the Time Traveller (he is never named, but in the early version he is Dr Nebo-gipfel, so perhaps it's just as well) invokes the fourth dimension to explain why such a machine is possible:

'But wait a moment. Can an instantaneous cube exist?'

'Don't follow you,' said Filby.

'Can a cube that does not last for any time at all, have a real existence?'

Filby became pensive.

'Clearly,' the Time Traveller proceeded, 'any real body must have extension in four directions: it must have Length, Breadth, Thickness, and – Duration . . .

' . . . There are really four dimensions, three which we call the three planes of Space, and a fourth, Time. There is, however, a tendency to draw an unreal distinction between the former three dimensions and the latter, because it happens that our consciousness moves intermittently in one direction along the latter from the beginning to the end of our lives . . .

' . . . But some philosophical people have been asking why three dimensions particularly – why not another direction at right angles to the three? – and have even tried to construct a Four-Dimensional geometry. Professor Simon Newcomb was expounding this to the New York Mathematical Society only a month or so ago.'

The notion of time as a fourth dimension was becoming common scientific currency in the late Victorian era. The mathematicians had started it, by wondering what a dimension was, and deciding that it need not be a direction in real space. A dimension was just a quantity that could be varied, and the number of dimensions was the largest number of such quantities that could all be varied independently. Thus the Discworld thaum, the basic particle of magic, is actually composed of resons, which come in at least five flavours: up, down, sideways, sex appeal, and peppermint. The thaum is therefore at least five-dimensional, assuming that up and down are independent, which is likely because it's quantum.

In the 1700s the foundling mathematician Jean le Rond D'Alembert (his middle name is that of the church where he was abandoned as a baby) suggested thinking of time as a fourth dimension in an article in the *Reasoned Encyclopaedia or Dictionary of Sciences, Arts, and Crafts*. Another mathematician, Joseph-Louis Lagrange, used time as a fourth dimension in his *Analytical Mechanics* of 1788, and his *Theory of Analytic Functions* of 1797 explicitly states: 'We may regard mechanics as a geometry of four dimensions.'

It took a while for the idea to sink in, but by Victorian times mathematicians were routinely combining space and time into a single entity. They didn't (yet) call it spacetime, but they could see that it had four dimensions: three of space plus one of time. Journalists and the lay public soon began to refer to time as *the* fourth dimension, because they couldn't think of another one, and to talk as if scientists had been looking for it for ages and had just found it. Newcomb wrote about the science of four-dimensional space from 1877, and spoke about it to the New York Mathematical Society in 1893.

Wells's mention of Newcomb suggests a link to one of the more colourful members of Victorian society, the writer Charles Howard Hinton. Hinton's primary claim to fame is his enthusiastic promotion of 'the' fourth dimension. He was a talented mathematician with a genuine flair for four-dimensional geometry, and in 1880 he published 'What is the Fourth Dimension?' in the *Dublin University Magazine*, which was reprinted in the *Cheltenham Ladies' Gazette* a year later. In 1884 it reappeared as a pamphlet with the subtitle 'Ghosts Explained'. Hinton, something of a mystic, related the fourth dimension to pseudoscientific topics ranging from ghosts to the after-life. A ghost can easily appear from, and disappear along, a fourth dimension, for instance, just as a coin can appear on, and disappear from, a tabletop, by moving along 'the' third dimension.

Charles Hinton was influenced by the unorthodox views of his sur-geon father James, a collaborator of Havelock Ellis, who outraged Victorian society with his studies of human sexual behaviour. Hinton the elder advocated free love and polygamy, and eventually headed a cult. Hinton the younger also had an eventful private life: in 1886 he fled to Japan, having been convicted of bigamy at the Old Bailey. In 1893 he left Japan to become a mathematics instructor at Princeton University, where he invented a baseball-pitching machine that used gunpowder to propel the balls, like a cannon. After several accidents the device was abandoned and Hinton lost his job, but his continu-ing efforts to promote the fourth dimension were more successful. He wrote about it in popular magazines like *Harper's Weekly*, *McClure's*, and *Science*. He died suddenly of a cerebral haemorrhage in 1907, at the annual dinner of the Society of Philanthropic Enquiry, having just completed a toast to female philosophers.

It was probably Hinton who put Wells on to the narrative possi-bilities of time as the fourth dimension. The evidence is indirect but compelling. Newcomb definitely knew Hinton: he once got Hinton a job. We don't know whether Wells ever met Hinton, but there is circumstantial evidence of a close connection. For example, the term 'scientific romance' was coined by Hinton in titles of his collected speculative essays in 1884 and 1886, and Wells later used the same phrase to describe his own stories. Moreover, Wells was a regular reader of *Nature*, which reviewed Hinton's first series of *Scientific Romances* (favourably) in 1885 and summarised some of his ideas on the fourth dimension.

In all likelihood, Hinton was also partially responsible for another

Victorian transdimensional saga, Edwin A. Abbott's *Flatland* of 1884. The tale is about A. Square, who lives in the Euclidean plane, a two-dimensional society of triangles, hexagons and circles, and doesn't believe in the third dimension until a passing sphere drops him in it. By analogy, Victorians who didn't believe in the fourth dimension were equally blinkered. A subtext is a satire on Victorian treatment of women and the poor. Many of Abbott's ingredients closely resemble elements found in Hinton's stories.*

Most of the physics of time travel is general relativity, with a dash of quantum mechanics. As far as the wizards of Unseen University are concerned, all this stuff is 'quantum' – a universal intellectual get-out-of-jail card – so you can use it to explain virtually anything, however bizarre. Indeed, the more bizarre, the better. You're about to get a solid dose of quantum in Chapter 8. Here we'll set things up by providing a quick primer on Einstein's theories of relativity: special and general.

As we explained in *The Science of Discworld*, 'relativity' is a silly name. It should have been 'absolutivity'. The whole point of special relativity is *not* that 'everything is relative', but that one thing – the speed of light – is unexpectedly *absolute*. Shine a torch from a moving car, says Einstein: the extra speed of the car will have no effect on the speed of the light. This contrasts dramatically with old-fashioned Newtonian physics, where the light from a moving torch would go faster, acquiring the speed of the car in addition to its own inherent speed. If you throw a ball from a moving car, that's what happens. If you throw light, it should do the same, but it doesn't. Despite the shock to human intuition, experiments show that Roundworld really does behave relativistically. We don't notice because the difference between Newtonian and Einsteinian

* See *The Annotated Flatland* by Edwin A. Abbott and Ian Stewart (Basic Books, 2002).

physics becomes noticeable only when speeds get close to that of light.

Special relativity was inevitable; scientists were bound to think of it. Its seeds were already sown in 1873 when James Clerk Maxwell wrote down his equations for electromagnetism. Those equations make sense in a 'moving frame' – when observations are made by a moving observer – only if the speed of light is *absolute*. Several mathematicians, among them Henri Poincaré and Hermann Minkowski, realised this and anticipated Einstein on a mathematical level, but it was Einstein who first took the ideas seriously as physics. As he pointed out in 1905, the physical consequences are bizarre. Objects shrink as they approach the speed of light, time slows to a crawl, and mass becomes infinite. Nothing (well, no *thing*) can travel faster than light, and mass can turn into energy.

In 1908 Minkowski found a simple way to visualise relativistic physics, now called Minkowski spacetime. In Newtonian physics, space has three fixed coordinates – left/right, front/back, up/down. Space and time were thought to be independent. But in the relativistic setting, Minkowski treated time as an extra coordinate in its own right. A fourth coordinate, a fourth independent direction . . . a fourth *dimension*. Three-dimensional space became four-dimensional spacetime. But Minkowski's treatment of time added a new twist to the old idea of D'Alembert and Lagrange. Time could, to some extent, be swapped with space. Time, like space, became geometrical.

We can see this in the relativistic treatment of a moving particle. In Newtonian physics, the particle sits in space, and as time passes, it moves around. Newtonian physics views a moving particle the way we view a movie. Relativity, though, views a moving particle as the sequence of still frames that make up that movie. This lends relativity an explicit air of determinism. The movie frames already exist before you run the movie. Past, present and future are already *there*. As time flows, and the movie runs, we discover what fate has in store for us – but fate is really *destiny*, inevitable, inescapable. Yes – the movie frames could perhaps come into existence one by one, with the newest one being the present, but it's not possible to do this consistently for every observer.

Relativistic spacetime = geometric narrativium.

Geometrically, a moving point traces out a *curve*. Think of the particle as the point of a pencil, and spacetime as a sheet of paper, with space running horizontally and time vertically. As the pencil moves, it leaves a line behind on the paper. So, as a particle moves, it traces out a curve in spacetime called its world-line. If the particle moves at a constant speed, the world-line is straight. Particles that move very slowly cover a small amount of space in a lot of time, so their world-lines are close to the vertical; particles that move very fast cover a lot of space in very little time, so their world-lines are nearly horizontal. In between, running diagonally, are the world-lines of particles that cover a given amount of space in the same amount of time – measured in the right units. Those units are chosen to correspond via the speed of light – say years for time and light-years for space. What covers one light-year of space in one year of time? Light, of course. So diagonal world-lines correspond to particles of light – *photons* – or anything else that can move at the same speed.

Relativity forbids bodies that move faster than light. The world-lines that correspond to such bodies are called timelike curves, and the timelike curves passing through a given event form a cone, called its 'light cone'. Actually, this is like two cones stuck together at their sharp tips, one pointing forward, the other backward. The forward-pointing cone contains the future of the event, all the points in spacetime that it could possibly influence. The backward-pointing cone contains its past, the events that could possibly influence *it*. Everything else is forbidden territory, elsewheres and elsewhens that have no possible causal connections to the chosen event.

Minkowski spacetime is said to be 'flat' – it represents the motion of particles when no forces are acting on them. Forces change the motion, and the most important force is gravity. Einstein invented general relativity in order to incorporate gravity into special relativity. In Newtonian physics, gravity is a force: it pulls particles away from the straight lines that they would naturally follow if no force were acting. In general relativity, gravity is a geometric feature of the universe – a form of spacetime curvature.

In Minkowski spacetime, points represent events, which have a location in both space and time. The 'distance' between two events must capture how far apart they are in space, *and* how far apart they are in time. It turns out that the way to do this is, roughly speaking, to take the distance between them in space and *subtract* the distance between them in time. This quantity is called the *interval* between the two events. If, instead, you did what seems obvious and *added* the time-distance to the space-distance, then space and time would be on exactly the same physical footing. However, there are clear differences: free motion in space is easy, but free motion in time is not. Subtracting the time-difference reflects this distinction; mathematically it amounts to considering time as *imaginary* space – space multiplied by the square root of minus one. And it has a remarkable effect: if a particle travels with the speed of light, then the interval between any two events along its world-line is zero.

Think of a photon, a particle of light. It travels, of course, at the speed of light. As one year of time passes, it travels one light-year. The sum of 1 and 1 is 2, but that's not how you get the interval. The interval is the difference $1 - 1$, which is 0. So the interval is related to the apparent rate of passage of time for a moving observer. The faster an object moves, the slower time on it appears to pass. This effect is called *time dilation*. As you travel closer and closer to the speed of light, the passage of time, as you experience it, slows down. If you could travel *at* the speed of light, time would be frozen. No time passes on a photon.

In Newtonian physics, particles that move when no forces are acting follow straight lines. Straight lines minimise the distance between points. In relativistic physics, freely moving particles minimise the interval, and follow *geodesics*. Finally, gravity is incorporated, not as an extra force, but as a distortion of the structure of spacetime, which changes the size of the interval and alters the shapes of geodesics. This variable interval between nearby events is called the *metric* of spacetime.

The usual image is to say that spacetime becomes 'curved', though this term is easily misinterpreted. In particular, it doesn't have to be curved *round* anything else. The curvature is interpreted physically as the force of gravity, and it causes light cones to deform.

One result is 'gravitational lensing', the bending of light by massive objects, which Einstein discovered in 1911 and published in 1915. He predicted that gravity should bend light by twice the amount that Newton's Laws imply. In 1919 this prediction was confirmed, when Sir Arthur Stanley Eddington led an expedition to observe a total eclipse of the Sun in West Africa. Andrew Crommelin of Greenwich Observatory led a second expedition to Brazil. The expeditions observed stars near the edge of the Sun during the eclipse, when their light would not be swamped by the Sun's much brighter light. They found slight displacements of the stars' apparent positions, consistent with Einstein's predictions. Overjoyed, Einstein sent his mum a postcard: 'Dear Mother, joyous news today . . . the English expeditions have actually demonstrated the deflection of light from the Sun.' the *Times* ran the headline: REVOLUTION IN SCIENCE. NEW THEORY OF THE UNIVERSE. NEWTONIAN IDEAS OVERTHROWN. Halfway down the second column was a subheading: SPACE 'WARPED'. Einstein became an overnight celebrity.

It would be churlish to mention that to modern eyes the observational data are decidedly dodgy – there might be some bending, and then again, there might not. So we won't. Anyway, later, better experiments confirmed Einstein's prediction. Some distant quasars produce multiple images when an intervening galaxy acts like a lens and bends their light, to create a cosmic mirage.

The metric of spacetime is not flat.

Instead, near a star, spacetime takes the form of a curved surface that bends to create a circular 'valley' in which the star sits. Light follows geodesics across the surface, and is 'pulled down' into the hole, because that path provides a short cut. Particles moving in spacetime at sublight speeds behave in the same way; they no longer follow straight lines, but are deflected towards the star, whence the Newtonian picture of a gravitational force.

Far from the star, this spacetime is very close indeed to Minkowski spacetime; that is, the gravitational effect falls off rapidly and soon becomes negligible. Spacetimes that look like Minkowski spacetime at large distances are said to be 'asymptotically flat'. Remember that term: it's important for making time machines. Most of our own universe is asymptotically flat, because massive bodies such as stars are scattered very thinly.

When setting up a spacetime, you can't just bend things any way you like. The metric must obey the Einstein equations, which relate the motion of freely moving particles to the degree of distortion away from flat spacetime.

We've said a lot about how space and time behave, but what are they? To be honest, we haven't a clue. The one thing we're sure of is that appearances can be deceptive.

Tick.

Some physicists take that principle to extremes. Julian Barbour, in *The End of Time*, argues that from a quantum-mechanical point of view, time does not exist.

Ti—

In 1999, writing in *New Scientist*, he explained the idea roughly this way. At any instant, the state of every particle in the entire universe can be represented by a single point in a gigantic phase space, which he calls Platonia. Barbour and his colleague Bruno Bertotti found out how to make conventional physics work in Platonia. As time passes, the configuration of all particles in the universe is represented in Platonia as a moving point, so it traces out a path, just like a relativistic world-line. A Platonian deity could bring the points of that path into existence sequentially, and the particles would move, and time would seem to flow.

Quantum Platonia, however, is a much stranger place. Here, 'quantum mechanics kills time', as Barbour puts it. A quantum particle is not a point, but a fuzzy probability cloud. A quantum state of the universe is a fuzzy cloud in Platonia. The 'size' of that cloud, relative to that of Platonia itself, represents the probability that the universe is in one of the states that comprise the cloud. So we have to endow Platonia with a 'probability mist', whose density in any given region determines how probable it is for a cloud to occupy that region.

But, says Barbour, 'there cannot be probabilities at different times, because Platonia itself is timeless. There can only be once-and-for-all probabilities for each possible configuration.' There is only one probability mist, and it is always the same. In this set-up, time is an illusion. The future is not determined by the present – not because of the role of chance, but because there is no such thing as future or present.

By analogy, think of the childhood game of snakes and ladders. At each roll of the dice, players move their counters from square to square on a board; traditionally there are a hundred squares. Some are linked by ladders, and if you land at the bottom you immediately rise to the top; others are linked by snakes, and if you land at the top you immediately fall to the bottom. Whoever reaches the final square first wins.

To simplify the description, imagine someone playing solo snakes and ladders, so that there is only one counter on the board. Then at any instant, the 'state' of the game is determined by a single square: whichever one is currently occupied by the counter. In this analogy, the board itself becomes the phase space, our analogue of Platonia.

The counter represents the entire universe. As the counter hops around, according to the rules of the game, the state of the 'universe' changes. The path that the counter follows – the list of squares that it successively occupies – is analogous to the world-line of the universe. In this interpretation, time does exist, because each successive move of the counter corresponds to one tick of the cosmic clock.

Quantum snakes and ladders is very different. The board is the same, but now all that matters is the probability with which the counter occupies any given square – not just at one stage of the game, but overall. For instance, the probability of being on the first square, at some stage in the game, is 1, because you always start there. The probability of being on the second square is 1/6, because the only way to get there is to throw a 1 with the dice on your first throw. And so on. Once we have calculated all these probabilities, we can forget about the rules of the game and the concept of a 'move'. Now only the probabilities remain. This is the quantum version of the game, and it has no explicit moves, only probabilities. Since there are no moves, there is no notion of the 'next' move, and no sensible concept of time.

Our universe, Barbour tells us, is a quantum one, so it is like quantum snakes and ladders, and 'time' is a meaningless concept. So why do we naive humans imagine that time flows; that the universe (at least, the bit near us) passes through a linear sequence of changes?

To Barbour, the apparent flow of time is an illusion. He suggests that Platonian configurations which have high probability must contain within them 'an appearance of history'. They will look *as though* they had a past. It's a bit like the philosophers' old chestnut: maybe the universe is being created anew every instant (as in *Thief of Time*), but at each moment, it is created along with apparent records of a lengthy past history. Such apparently historical clouds in Platonia are called time capsules. Now, among those high-probability configurations we find the arrangement of neurons in a conscious brain. In other words, the universe itself is timeless, but our brains are time capsules, high-probability configurations, and these automatically come along with the *illusion* that they have had a past history.

It's a neat idea, if you like that sort of thing. But it hinges on Barbour's claim that Platonia must be timeless because 'there can only be once-and-for-all probabilities for each possible configuration'. This statement is remarkably reminiscent of one of Xeno's – sorry, Zeno's – paradoxes: the Arrow. Which, you recall, says that at each instant an arrow has a specific location, so it can't be moving. Analogously, Barbour tells us that at each instant (if such a thing could exist) Platonia must have a specific probability mist, and deduces that this mist can't change (so it doesn't).

What we have in mind as an alternative to Barbour's timeless probability mist is not a mist that changes as time passes, however. That would fall foul of the non-Newtonian relation between space and time; different parts of the mist would correspond to different times depending on who observed them. No, we're thinking of the mathematical resolution of the Arrow paradox, via Hamiltonian mechanics. Here, the state of a body is given by *two* quantities, position and momentum, instead of just position. Momentum is a 'hidden variable', observable only through its effect on subsequent positions, whereas position is something we can observe directly. We said: 'a body in a given position with zero momentum is not moving at that instant, whereas one in the same position with non-zero momentum *is* moving, even though instantaneously it stays in the same place'. Momentum encodes the next change of position, and it encodes it *now*. Its value now is not observable now, but it is (will be) observable. You just have to wait to find out what it was. Momentum is a 'hidden variable' that encodes *transitions* from one position to another.

Can we find an analogue of momentum in quantum snakes and ladders? Yes, we can. It is the overall probability of going from any given square to any other. These 'transition probabilities' depend only on the squares concerned, not on the time at which the move is made, so in Barbour's sense they are 'timeless'. But when you are on some given square, the transition probabilities tell you where your *next* move can lead, so you can reconstruct the possible sequences of moves, thereby putting time back into the physics.

For exactly the same reason, a single fixed probability mist is *not* the only statistical structure with which Platonia can be endowed. Platonia can also be equipped with transition probabilities between *pairs* of states. The result is to convert Platonia into what statisticians call a 'Markov chain', which is just like the list of transition probabilities for snakes and ladders, but more general. If Platonia is made into a Markov chain, each *sequence* of configurations gets its own probability. The most probable sequences are those that contain large numbers of highly probable states – these look oddly like Barbour's time capsules. So instead of single-state Platonia we get sequential-state Markovia, where the universe makes transitions through whole sequences of configurations, and the most likely transitions are the ones that provide a coherent history – narrativium.

This Markovian approach offers the prospect of bringing time back into existence in a Platonian universe. In fact, it's very similar to how Susan Sto Helit and Ronnie Soak managed to operate in the cracks between the instants, in *Thief of Time*.

Tick.

SEVEN

THE FISH
IS OFF

TWO HOURS LATER A SINGLE sheet of paper slid off Hex's writing table. Ponder picked it up.

'There are about ten points where we must intervene to ensure that *The Origin* is written,' he said.

'Well, that doesn't seem too bad,' said Ridcully. 'We got Shakespeare born, didn't we?* We just have to tinker.'

'These look a little more complicated,' said Ponder, doubtfully.

'But Hex can move us around,' said Ridcully. 'It could be fun, especially if something or someone is playing *les buggeurs risibles*. It could be educational, Mr Stibbons.'

'And they do really good beer, ' said the Dean. 'And the food was excellent. Remember that goose we had last time? I've seldom eaten better.'

'We will be setting out to save the world,' said Ridcully, severely. 'We will have other things on our minds!'

'But there will be mealtimes, yes?' said the Dean

Second Lunch and Mid-afternoon Snack went past almost unnoticed. Perhaps the wizards were already leaving space for goose . . .

*

* Yes, they did – in *The Science of Discworld II*.

It was turning out to be a long day. Easels had been set up around Hex. Paper was strewn across every table. The Librarian had practically built up a branch library in one corner, and was still fetching books from the distant reaches of L-Space.

And the wizards had changed their clothes, ready for hands-on intervention. There had barely been a discussion about it, not after the Dean had mentioned the goose. Hex had a great deal of control over the Globe, but when it came to the fine detail you needed to be hands-on, especially hands on cutlery. Hex had no hands. Besides, he'd explained at length, there was no such thing as absolute control, not in a fully functioning universe. There was just a variable amount of lack of control. In fact, Ponder thought, Hex was a Great Big Thing as far as Roundworld was concerned. Almost . . . godlike. But he still couldn't control *everything*. Even if you knew where every tiny spinning particle of stuff was, you couldn't know what it'd do next.

The wizards would have to go in. They could do that. They'd done it before. No trouble is too much if it saves some excellent chefs from extinction.

Clothing, at least, would not be a problem. Give or take the odd pointy hat and staff, the wizards would be able to walk the Roundworld streets without attracting a second glance.

'How do we look?' said the Archchancellor, as they reassembled.

'Very . . . Victorian,' said Ponder. 'Although technically, at the moment, very Georgian. Very . . . tweedy, anyway. Are you totally happy with the bishop look, Dean?'

'Isn't that appropriate for the time?' said the Dean, looking worried. 'We looked through the book on costumes and I thought . . .' His voice trailed off. 'It's the mitre, isn't it . . .'

'And the crozier,' said Ponder.

'I wanted to fit in, you see.'

'In a cathedral, yes. I'm afraid it's plain black suit with gaiters for street wear. However, you can do anything you like with your beard and you can wear hats a small child could stand up inside. But on the streets, bishops are quite dull.'

'Where's the fun in that?' said the Dean, sulkily.

Ponder turned to Rincewind.

'As for you, Rincewind, can I ask why you are wearing nothing but a loincloth and a pointy hat?'

'Ah, well, you see, if you don't know what you're getting into, naked always works,' said Rincewind. 'It's the all-purpose suit. At home in every culture. Admittedly you sometimes get—'

'In tweed, that man!' barked Ridcully. 'And no pointy hat!' Against a background of grumbling he turned then to the Librarian. 'And as for you, sir . . . a suit too. And a stovepipe hat. You need the height!'

'Ook!' said the Librarian.

'I *am* the Archchancellor, sir! I insist! And a false beard, I think. False eyebrows, too. Let Mr Darwin be your model here! These Victorians were very civilised people! Hair everywhere! Keep the knuckling to a minimum and they'll make you Prime Minister! Very well, gentlemen. Back here in half an hour!

The wizards assembled. A circle of white light appeared on the floor. They stepped inside, there was a change in the sounds made by Hex, and they vanished.

They landed knee-deep in the mire of a peat bog, causing bubbles of foul air to burst around them.

'Mr Stibbons!' Ridcully bellowed.

'Sorry, sir, sorry,' said Ponder quickly. 'Hex, raise us by two feet, please.'

'Yes, but we're still soaked,' grumbled the Dean, as they floated up in the air. 'You seem to have, ah, "mucked up", Mr Stibbons!'

'No, sir, I wanted to show you a Charles Darwin in the wild,' said Ponder. 'Here he comes now . . .'

A large and energetic young man bounded out of the weeds and went to clear a black pool with a vaulting pole. The pole immediately sank one-third of its length into the sucking ground and its athletic owner sailed off into the mud. He came up holding a small water plant. Oblivious of the noisome bubbling around him, he waved the plant triumphantly at some distant companions, pulled his pole out of the peat with some effort, and splashed away.

'Did he see us?' said Rincewind.

'No, not yet. That's *young* Darwin,' said Ponder. 'Very keen on collecting all sorts of wildlife. Collecting was enormous popular among the English of this century. Bones, shells, butterflies, birds, other people's countries . . . all sorts of things.'

'Man after my own heart!' said Ridcully, cheerfully. 'I had the best pressed lizard collection ever when I was that age!'

'Can't see a beagle anywhere, though,' said Rincewind, gloomily. He got edgy in the absence of his hat, and tried to stand under things.

The Chair of Indefinite Studies looked up from the thaumometer in his hand.

'No magic disturbance, no nothing,' he said, looking around at the marshes. 'Is Hex *sure*? The only strange thing here is us.'

'Let's get started, shall we?' said Ridcully. 'Where to next?'

'Hex, move us to London, will you?' said Ponder. 'Location 7.'

The wizards didn't apparently move, but the landscape around them wavered and changed.

It became an alleyway. There were a lot of street noises nearby.

'I'm sure you all read the briefing I prepared this morning.' said Ponder, brightly.

'Are you also sure we've not back in Ankh-Morpork?' said Ridcully loudly. 'I'd swear I can smell the river!'

'Ah, then perhaps I'd better just remind you of the *important* points,' said Ponder wearily. 'The list of major things that might impede Darwin's progress—'

'I remember about the giant squid,' Rincewind volunteered.

'Hex can handle the giant squid,' said Ponder.

'Oh, shame. I was looking forward to that,' said Ridcully.

'No, sir,' said Ponder, as patiently as possible. '*We* have to deal with people. Remember? We agreed last time it's not ethical to leave that to Hex. Remember the rain of fat women?'*

'That never actually *happened*,' said the Lecturer in Recent Runes, wistfully.

'Quite so,' said Ridcully, firmly. 'And just as well. Lead on, Mr Stibbons.'

'So much to do, so much to do,' muttered Ponder, leafing through the paperwork. 'I suppose we'd better do things in order . . . so first, we must see that Mr Habbakuk Souser's cook throws away the fish.'

It was a scullery boy who opened the back door, in a street of quite prosperous-looking houses. Ponder Stibbons raised his very tall hat.

'We wish to see – ' he consulted the clipboard ' – Mrs Boddy,' he said. 'She is the cook here, I believe? Tell here we are the Committee for Public Sanitation, and the matter is urgent, so look sharp about it!'

'I hope you know what you're doing, Stibbons,' hissed Ridcully as the boy scurried away.

'Totally, Archchancellor. Hex says the line of causality is – ah, Mrs Boddy?'

This was to a skinny, worried woman who was advancing on them from the dim interior, wiping her hands on her apron.

'I am, sir,' said the cook. 'The boy said you gentlemen was Hygienic?'

'Mrs Boddy, you had some fish delivered this morning?' said Ponder, sternly.

* A rare meteorological phenomenon discussed briefly in *The Science of Discworld II*.

'Yes sir. Nice piece o' hake.' Sudden uncertainty seized her features. 'Er . . . that was all right, wasn't it?'

'Alas it was not, Mrs Boddy!' said Ponder. 'We have just come from the fishmonger. All his hake is completely off. We have had many complaints. Some of them were from next of kin, Mrs Boddy!'

'Oh, what shall we do to be saved!' the cook burst out. 'I've got it cookin'! It smelled all right, sir!'

'Thankfully, then, there is no harm done,' said Ponder.

'Shall I give it to the cat?'

'Do you *like* the cat?' said Ponder. 'No, wrap it in some paper and bring it out to us right now! I'm sure Mr Souser will understand when you give him some of the cold ham from yesterday.'

'Yessir!' The cook scurried away, and returned shortly with a parcel of very hot, very damp fish. Ponder grabbed it from her and thrust it into Rincewind's arms.

'Scour the pan thoroughly, Mrs Boddy!' said Ponder, as Rincewind tried to juggle hake. 'Gentlemen, we must hurry!'

He started to walk very fast towards the end of the street, the wizards jogging along behind him, and turned sharply into an alleyway just ahead of a shout of 'Sir? Sir? How did you know about the cold ham?'

'Location 9, Hex,' said Ponder. 'And remove the fish, please!'

'Was all that about?' said Ridcully. 'Why did we take that poor woman's fish?'

Rincewind said 'ow!' as the fish disappeared.

'Mr Souser will travel, er, tomorrow to meet some businessmen,' said Ponder, as a circle formed on the ground around the wizards. 'One of them will be a man called Josiah Wedgwood, a famous industrialist. Mr Souser will tell him about his son James, who is currently working with the Navy. It has made a man of him, Mr Souser will say. Mr Wedgwood will listen with interest, and form the opinion that the adventure of a long sea voyage in respectable company may well be of benefit to a young man on the verge of adult life. At least, he will now. If Mr Souser had eaten that fish, he would have been too ill to travel tomorrow.'

'Well, that's good news for Mr Souser, but what's it got to do with us?' said the Dean.

'Mr Wedgwood is Charles Darwin's uncle,' said Ponder, as the air wavered. 'He will have an influence on his nephew's career. And now for our next call . . .'

'Good morning! Mrs Nightingale?'

'Yes?' said the woman, as if she was now doubting it. She took in the group of people in front of her, her eye resting on the very bearded one whose knuckles touched the ground. Beside her, the housemaid who'd opened the door looked on nervously.

'My name is Mr Stibbons, Mrs Nightingale. I am the secretary of The Mission to Deep Sea Voyagers, a charitable organisation. I believe Mr Nightingale is shortly to embark on a perilous mission to the storm-tossed, current-mazed, ship-eating giant-squid infested waters of the South Americas?'

The woman's gaze tore itself away from the Librarian and her eyes narrowed.

'He never said anything to *me* about giant squid,' she said.

'Indeed? I'm very sorry to hear that, Mrs Nightingale. Brother Bookmeister here,' Ponder patted the Librarian on the shoulder, 'would tell you about them himself were it not that the dire experience quite robbed him of the power of speech.'

'Ook!' said Brother Bookmeister plaintively.

'Really?' said the woman, setting her jaw firmly. 'Would you gentlemen care to step into the parlour?'

'Well, the biscuits were nice,' said the Dean, as the wizards strolled out into the street half an hour later. 'And now, Stibbons, would you care to tell us what all that was about?'

'Gladly, Dean, and may I say your story about the sea snake was very useful?' said Ponder. 'But Rincewind, that tale about the killer flying fish was rather over the top, I thought.'

'I didn't make it up!' Rincewind said. 'They had teeth on them like—'

'Well, anyway . . . Darwin was the second choice for the post on the *Beagle*,' said Ponder. 'Mr Nightingale was the captain's initial choice. History will record that after his wife's pleading he declined the offer. This he will do within about five minutes of when he gets home tonight.'

'Another fine ruse?' said Ridcully.

'I'm rather pleased with it, as a matter of fact,' said Ponder.

'Hmm,' said Ridcully. Cunning in younger wizards is not automatically applauded in their elders. 'Very clever, Stibbons. You are a wizard to watch.'

'Thank you, sir. My next question is: does anyone here know anything about shipbuilding? Well, perhaps that won't be necessary. Hex, take us to Portsmouth, please. The *Beagle* is being refitted. You will need to be naval inspectors which, ahaha, I'm sure you'll be good at. In fact you will be the most observant inspectors there have ever been. Location 3, please, Hex.'

EIGHT

FORWARD TO
THE PAST

 WELL, THE WIZARDS HAVE MADE a good start. And with the might of Hex behind them, the wizards can travel at will along the Roundworld timeline. We're happy for them to do that, in a fictional context – but could we do the same thing, in a factual one?

To answer that, we must decide what a time machine looks like within the framework of general relativity. Then we can talk about building one.

Travel into the future is easy: wait. It's getting back that's hard. A time machine lets a particle or object return to its own past, so its world-line, a timelike curve, must close into a loop. So a time machine is just a *closed timelike curve*, abbreviated to CTC. Instead of asking, 'Is time travel possible?' we ask, 'Can CTCs exist?'

In flat Minkowski spacetime, they can't. Forward and backward light cones – the future and past of an event – never intersect (except at the point itself, which we discount). If you head off across a flat plane, never deviating more than 45° from due north, you can never sneak up on yourself from the south.

But forward and backward light cones can intersect in other types of spacetime. The first person to notice this was Kurt Gödel, better known for his fundamental work in mathematical logic. In 1949 he worked out the relativistic mathematics of a rotating universe, and discovered that the past and future of every point intersect. Start wherever and whenever you like, travel into your future, and you'll end up in your own past. However, observations indicate that the universe is not rotating, and spinning up a stationary universe (especially from inside) doesn't look like a plausible way to make a time machine. Though, if the wizards were to give Roundworld a twirl . . .

The simplest example of future meeting past arises if you take Minkowski spacetime and roll it up along the 'vertical' time direction to form a cylinder. Then the time coordinate becomes cyclic, as in Hindu mythology, where Brahma recreates the universe every *kalpa*, a period of 4.32 billion years. Although a cylinder looks curved, the corresponding spacetime is *not* actually curved – not in the gravitational sense. When you roll up a sheet of paper into a cylinder, it doesn't *distort*. You can flatten it out again and the paper is not folded or wrinkled. An ant that is confined purely to the surface won't notice that spacetime has been bent, because distances *on* the surface haven't changed. In short the local metric doesn't change. What changes is the global geometry of spacetime, its overall *topology*.

Rolling up Minkowski spacetime is an example of a powerful mathematical trick for building new spacetimes out of old ones: cut-and-paste. If you can cut pieces out of known spacetimes, and glue them together without distorting their metrics, then the result is also a possible spacetime. We say 'distorting the metric' rather than 'bending', for exactly the reason that we say that rolled-up Minkowski spacetime is *not* curved. We're talking about intrinsic curvature, as experienced by a creature that lives in the spacetime, not about apparent curvature as seen by some external viewer.

The rolled-up version of Minkowski spacetime is a very simple way to prove that spacetimes that obey the Einstein equations *can* possess CTCs – and thus that time travel is not inconsistent with currently known physics. But that doesn't imply that time travel is *possible*. There is a very important distinction between what is mathematically possible and what is physically feasible.

A spacetime is mathematically possible if it obeys the Einstein equations. It is physically feasible if it can exist, or could be created, as part of our own universe or an add-on. There's no very good reason to suppose that rolled-up Minkowski spacetime is physically feasible: certainly it would be hard to refashion the universe in that form if it wasn't already endowed with cyclic time, and right now very few people (other than Hindus) think that it is. The search for spacetimes that possess CTCs *and* have plausible physics is a search for more plausible topologies. There are many mathematically possible topologies, but, as with the Irishman giving directions, you can't get to all of them from here.

However, you can get to some remarkably interesting ones. All you need is black hole engineering. Oh, and white holes, too. And negative energy. And –

One step at a time. Black holes first. They were first predicted in classical Newtonian mechanics, where there is no limit to the speed of a moving object. Particles can escape from an attracting mass, however strong its gravitational field, by moving faster than the appropriate 'escape velocity'. For the Earth, this is 7 miles per second (11 kps), and for the Sun, it is 26 miles per second (41 kps). In an article presented to the Royal Society in 1783, John Michell observed that the concept of escape velocity, combined with a finite speed of light, implies that sufficiently massive objects cannot emit light at all – because the speed of light will be lower than the escape velocity. In 1796 Pierre Simon de Laplace repeated these observations in his *Exposition of the System of the World*. Both of them imagined that the universe might be littered with huge bodies, bigger than stars, but totally dark.

They were a century ahead of their time.

In 1915 Karl Schwarzschild took the first step towards answering the relativistic version of the same question, when he solved the

Einstein equations for the gravitational field around a massive sphere in a vacuum. His solution behaved very strangely at a critical distance from the centre of the sphere, now called the Schwarzschild radius. It is equal to the mass of the star, multiplied by the square of the speed of light, multiplied by twice the gravitational constant, if you must know.

The Schwarzschild radius for the Sun is 1.2 miles (2 km), and for the Earth 0.4 inches (1 cm) – both buried inaccessibly deep where they can't cause trouble. So it wasn't entirely clear how significant the strange mathematical behaviour was . . . or even what it meant.

What would happen to a star that is so dense that it lies inside its own Schwarzschild radius?

In 1939 Robert Oppenheimer and Hartland Snyder showed that it would collapse under its own gravitational attraction. Indeed a whole portion of spacetime would collapse to form a region from which no matter, not even light, could escape. This was the birth of an exciting new physical concept. In 1967 John Archibald Wheeler coined the term *black hole*, and the new concept was christened.

How does a black hole develop as time passes? An initial clump of matter shrinks to the Schwarzschild radius, and then continues to shrink until, after a finite time, all the mass has collapsed to a single point, called a singularity. From outside, though, we can't observe the singularity: it lies beyond the 'event horizon' at the Schwarzschild radius, which separates the observable region, from which light can escape, and the unobservable region where the light is trapped.

If you were to watch a black hole collapse from outside, you would see the star shrinking towards the Schwarzschild radius, but you'd never see it get there. As it shrinks, its speed of collapse as seen from outside approaches that of light, and relativistic time-dilation implies that the entire collapse takes infinitely long when seen by an outside observer. The light from the star would shift deeper and deeper into the red end of the spectrum. The name should be 'red hole'.

Black holes are ideal for spacetime engineering. You can cut-and-paste a black hole into any universe that has asymptotically flat regions, such as our own.* This makes black hole topology physically plausible in our universe. Indeed, the scenario of gravitational collapse makes it even more plausible: you just have to start with a big enough concentration of matter, such as a neutron star or the centre of a galaxy. A technologically advanced society could *build* black holes.

A black hole doesn't possess CTCs, though, so we haven't achieved time travel. Yet. However, we're getting close. The next step uses the time-reversibility of Einstein's equations: to every solution there corresponds another that is just the same, except that time runs backwards. The time reversal of a black hole is called a white hole. A black hole's event horizon is a barrier from which no particle can escape; a white hole's event horizon is one into which no particle can fall, but from which particles may emerge at any moment. So, seen from the outside, a white hole would appear as the sudden explosion of a star's worth of matter, coming from a time-reversed event horizon.

White holes may seem rather strange. It makes sense for an initial concentration of matter to collapse, if it is dense enough, and thus to form a black hole; but why should the singularity inside a white hole suddenly decide to spew forth a star, having remained unchanged since the dawn of time? Perhaps because time runs backwards inside a white hole, so causality runs from future to past? Let's just agree that white holes are a mathematical possibility, and notice that they too are asymptotically flat. So if you knew how to make one, you could glue it neatly into your own universe, too.

Not only that: you can glue a black hole and a white hole together. Cut them along their event horizons, and paste along these two horizons. The result is a sort of tube. Matter can pass through the tube

---

* This is a mathematician's way of saying that you can put a black hole anywhere you want. (Or, like a gorilla in a Mini, it can go anywhere *it* wants.)

in one direction only: into the black hole and out of the white. It's a kind of matter-valve. The passage through the valve follows a time-like curve, because material particles can indeed traverse it.

Both ends of the tube can be glued into any asymptotically flat region of any spacetime. You could glue one end into our universe, and the other end into somebody else's; or you could glue both ends into ours – anywhere you like except near a concentration of matter. Now you've got a wormhole. The distance through the wormhole is very short, whereas that between the two openings, across normal spacetime, can be as big as you like.

A wormhole is a short cut through the universe.

But that's matter-transmission, not time travel.

Never mind: we're nearly there.

The key to wormhole time travel is the notorious twin paradox, pointed out by the physicist Paul Langevin in 1911. Recall that in relativity, time passes more slowly the faster you go, and stops altogether at the speed of light. This effect is known as time dilation. We quote from *The Science of Discworld*:

> Suppose that Rosencrantz and Guildenstern are born on Earth on the same day. Rosencrantz stays there all his life, while Guildenstern travels away at nearly lightspeed, and then turns round and comes home again. Because of time dilation, only one year (say) has passed for Guildenstern, whereas 40 years have gone by for Rosencrantz. So Guildenstern is now 39 years younger than his twin brother.

It's called a paradox because there seems to be a puzzle: from Guildenstern's frame of reference, it is Rosencrantz who has whizzed off at near-lightspeed. Surely, by the same token, Rosencrantz should be 39 years younger, not Guildenstern? But the apparent symmetry is fallacious. Guildenstern's frame of reference is subject to acceleration and deceleration, especially when he turns round to head for home; Rosencrantz's isn't. In relativity, accelerations make a big difference.

In 1988 Michael Morris, Kip Thorne, and Ulvi Yurtsever realised that combining a wormhole with the twin paradox yields a CTC. The idea is to leave the white end of the wormhole fixed, and to zigzag the black one back and forth at just below the speed of light. As the black end zigzags, time dilation comes into play, and time passes more slowly for an observer moving with that end. Think about world-lines that join the two wormholes through normal space, so that the time experienced by observers at each end are the same. At first those lines are almost horizontal, so they are not timelike, and it is not possible for material particles to proceed along them. But as time passes, the line gets closer to the vertical, and eventually it becomes timelike. Once this 'time barrier' is crossed, you can travel from the white end of the wormhole to the black through normal space – following a timelike curve. Because the wormhole is a short cut, you can do so in a very short period of time, effectively travelling instantly from the black end to the corresponding white one. This is the same place as your starting point, but in the past.

You've travelled in time.

By waiting, you can close the path into a CTC and end up at the same place and time that you started from. Not back to the future, but forward to the past. The further into the future your starting point is, the further back in time you can travel from that point. But there's one disadvantage of this method: you can never travel back past the time barrier, and that occurs some time after you build the wormholes. No hope of going back to hunt dinosaurs. Or to tread on Cretaceous butterflies.

Could we really make one of these devices? Could we really get through the wormhole?

In 1966 Robert Geroch discovered a theoretical way to warp space-time, smoothly, without tearing it, to create a wormhole. There's a snag, though: at one stage in the construction, time becomes so twisted that the wormhole turns into a temporary time machine, and equipment from late in the construction gets carried back to the beginning. The builders' tools might vanish into the past, just as they thought they had finished. Still, with the right work schedule, that might not matter. Perhaps a technologically advanced civilisation could build black and white holes, and move them around, by creating intense gravitational fields.

But building a wormhole is not the only obstacle. Keeping it open is another. The main trouble is the 'catflap effect': when you move a mass through a wormhole, the hole tends to shut on your tail. It turns out that in order to get through without getting your tail trapped, you have to travel faster than light, so that's no good. Any timelike path that starts at the wormhole entrance must run into the future singularity. There's no way to get across to the exit without exceeding the speed of light.

The traditional way round this difficulty is to thread the wormhole with 'exotic' matter, which exerts enormous negative pressure like a stretched spring. It is a form of negative energy, and is thus different from antimatter, which has positive energy. In quantum mechanics, a vacuum is not empty – it is a turbulent sea of particles, coming into being and disappearing again. Zero energy includes all these fluctuations, so you can get negative energy if you can calm the waves. One way to do this is the 'Casimir effect', discovered in 1948: if two metal plates are held very close together, then in between you find a negative energy state. This effect has been observed in experiments, but it's very weak. To get enough negative energy, you need galaxy-sized plates. Rigid ones, to maintain the gap.

Another possibility is a magnetic wormhole. In 1907 the geometer Tullio Levi-Civita proved that in general relativity a magnetic field can warp space. Magnetism has energy, energy is equivalent to mass, and mass is spatial curvature. Moreover, he found an exact mathematical solution to Einstein's field equations which he called 'magnetic gravity'. The trouble was, this effect could only be observed using a magnetic field one quintillion times as large as anything that could be obtained in a laboratory. The idea languished until 1995, when Claudio Maccone realised that Levi-Civita had actually come up with a magnetic wormhole. The stronger the magnetic field, the more tightly curved the wormhole mouth is. A wormhole whose magnetic field was the strength you can get in a laboratory would be enormous – about 150 light-years across. And you'd need laboratories everywhere along it. It's making a *small* wormhole that needs a gigantic magnetic field. Maccone suggested that the surface of a neutron star, where very strong magnetic fields can occur, might be a good place to look for magnetic wormholes. Why bother? Because a magnetic wormhole can stay open without any need for exotic matter.

A better solution, though, might be to employ a rotating black hole, which has a ring singularity, not a point one. Passers-by can go through the ring and miss the singularity. The mathematics of Einstein's equations tells us that a rotating black hole connects to infinitely many different regions of spacetime. One must be in our universe (assuming that we built the rotating black hole in our universe), but the others need not be. Beyond the ring singularity lie antigravity universes in which distances are negative and matter repels other matter. There are legal (slower-than-light) paths through the wormhole to any of its alternative exits. So, if we use a rotating black hole instead of a wormhole, and if we can find a way to tow its entrances and exits around at nearly lightspeed, we'll get a much more practical time machine – one that we can get through without running into the singularity.

*

There are other time machines based on the twin paradox, but all of them are limited by the speed of light. They would work better, and perhaps be easier to build and operate, if you could follow *Star Trek* and engage your warp drive, travelling faster than light.

But relativity forbids that, right?

Wrong.

Special relativity forbids that. General relativity, it turns out, permits it. The amazing thing is that the way it permits it turns out to be standard SF gobbledegook, invoked by innumerable writers who knew about relativistic limitations but still wanted their starships to travel faster than light. 'Relativity forbids *matter* travelling faster than light,' they would incant, 'but it doesn't forbid *space* travelling faster than light.' Put your starship in a region of space, and leave it stationary relative to that region. No violation of Einstein there. Now move the entire region of space, starship inside, with superluminal (faster-than-light) speed. Bingo!

Ha-ha, most amusing. Except . . .

In the context of general relativity, that's exactly what Miguel Alcubierre Moya came up with in 1994. He proved that there exist solutions of Einstein's field equations involving a local 'warping' of spacetime to form a mobile bubble. Space contracts ahead of the bubble and expands behind it. Put a starship inside the bubble, and it can 'surf' a gravitational wave, cocooned inside a static shell of local spacetime. The speed of the starship relative to the bubble is zero. Only the bubble's boundary moves, and that's just empty space.

The SF writers were right. There is no relativistic limit to the speed with which *space* can move.

Warp drives have the same drawback as wormholes. You need exotic matter to create the gravitational repulsion needed to distort spacetime in this unusual way. Other schemes for warp drives have been proposed, which allegedly overcome this obstacle, but they have their own drawbacks. Sergei Krasnikov noticed one awkward feature of Alcubierre's warp drive: the inside of the bubble becomes causally disconnected from the front edge. The starship's captain, inside the bubble, can't steer it, and she can't even turn it on or off. He proposed a different method, a 'superluminal highway'. On the outward trip, the starship travels below lightspeed and leaves a tube of distorted spacetime behind it. On the way back, it travels faster than light along the tube. The superluminal highway also needs negative energy; in fact, Ken Olum and others have proved that any type of warp drive does.

There are limits to the lifetime of any given amount of negative energy. For wormholes and warp drives these limits imply that such structures must either be very small, or else the region of negative energy must be extremely thin. For example, a wormhole whose mouth is three feet (1m) across must confine its negative energy to a band whose thickness is one millionth of the diameter of a proton. The total negative energy required would be equivalent to the total output (in positive energy) of 10 billion stars for one year. If the mouth were one light year across, then the thickness of the negative energy band would still be smaller than a proton, and now the negative energy requirement would be that of 10 quadrillion stars.

Warp drives, if anything, are worse. To travel at 10 times lightspeed (a mere *Star Trek* Warp Factor 2) the thickness of the bubble's wall must be 10^{-32} metres. If the starship is 200 yards (200m) long, the energy required to make the bubble has to be 10 billion times the mass of the known universe.

Engage.

Roundworld narrativium can sometimes be documented. When Ronald Mallett was ten years old his 33-year-old father died of heart failure, brought on by drinking and smoking. 'It completely devastated me,' he is reported to have said.* Soon after, he read Wells's *The Time*

---

* Michael Brooks, 'Time Twister', *New Scientist*, 19 May 2001, 27–9.

Machine. And he reasoned that 'If I could build a time machine, I might be able to warn him about what was going to happen.'

The childish idea faded, but the interest in time travel did not. As an adult, Mallett invented an entirely new type of time machine, one that uses bent light.

Morris and Thorne bent space to make a wormhole using matter. Mass *is* curved space. Levi-Civita bent space using magnetism. Magnetism has energy, energy *is* (so Einstein tells us) mass. Mallett prefers to bend space using light. Light, too, has energy. So it can act like mass. In 2000, he published a paper on the deformation of space by a circular beam of light. Then it hit him. If you can deform space, you ought to be able to deform time too. And his calculations showed that a ring of light could create a ring of time – a CTC.

With a Mallett bent-light time machine, you can walk into your past. A time traveller makes his or her way into the closed loop of light, space, and time. Walking round the loop has the same effect as moving backwards in time. The more times he walks round the loop, the further back he goes, tracing out a helical world-line. When he has gone sufficiently far into his past, he exits the loop. Easy.

Yes, but . . . we've been here before. It takes huge amounts of energy to make a circular beam of light.

That's true . . . unless you can slow the light down. A ring of really *slow* light, Discworld-speed, like that of sound on Roundworld, is much easier to make. The reason is that as light slows down, it *gains* inertia. This gives it more energy, and the warping effect is far greater for less effort on the part of the builder.

Relativity tells us that the speed of light is constant – *in a vacuum*. In other media, light slows down; this is why glass refracts light, for example. In the right medium, light can be slowed to walking pace, or even stopped altogether. Experiments by Lene Hau demonstrated this effect in 2001, using a medium known as a Bose-Einstein condensate. This is a curious, degenerate form of matter, occurring at temperatures near absolute zero; it consist of lots of atoms in exactly the same quantum state, forming a 'superfluid' with zero viscosity.

So maybe Wells's time traveller could have included some refrigeration equipment and a laser in his machine. But a Mallett bent-light time machine suffers from the same limitation as a wormhole one. You can't travel back to any time before the machine was constructed.

Wells was probably right to eliminate that encounter with a giant hippopotamus.

These are purely relativistic time machines, but the universe has quantum features too, and these should be taken into account. The search for a unification of relativity and quantum theory – respectably known as 'quantum gravity' and often derided as a Theory of Everything – has turned up a beautiful mathematical proposal, string theory. In this theory, instead of fundamental particles being points, they are vibrating multidimensional loops. The best-known version uses six-dimensional loops, so its model of spacetime is really ten-dimensional. Why has no one noticed? Perhaps because the extra six dimensions are curled up so tightly that no one has observed them – very possibly, *can* observe them. Or perhaps – the Irishman again – we can't go that way from here.

Many physicists hope that string theory, as well as unifying relativity and quantum mechanics, will also supply a proof of Hawking's chronology protection conjecture – that the universe conspires to keep events happening in the same temporal order. In this connection, there is a five-dimensional string-theoretic rotating black hole called a BMPV* black hole. If this rotates fast enough, it has CTCs *outside* the black hole region. Theoretically, you can build one from gravitational waves and esoteric string-theoretic gadgets called 'D-branes'.

* Jason Breckenridge, Rob Myers, Amanda Peet, and Cumrun Vafa.

And here we see a hint of Hawking's cosmological time cops. Lisa Dyson took a careful look at just what happens when you put the gravitational waves and D-branes together. Just as the black hole is within a gnat's whisker of turning into a time machine, the components stop collecting together in the same place. Instead, they form a shell of gravitons (hypothetical particles of gravity, analogous to photons for light). The D-branes are trapped inside the shell. The gravitons can't be persuaded to come any closer, and the BMPV can't be made to spin rapidly enough to create an accessible CTC.

The laws of physics won't let you put this kind of time machine together, unless some clever kind of scaffolding can be invented.

Quantum mechanics adds a new spin to the whole time-travel game. For a start, it may open up a way to create a wormhole. On the very tiny length scale of the quantum world, known as the Planck length (around 10^{-35} metres), spacetime is thought to be a quantum foam – a perpetually changing mass of tiny wormholes. Quantum foam is a kind of time machine. Time is slopping around inside the foam like spindrift bobbing on the ocean waves. You just have to harness it. An advanced civilisation might be able to use gravitational manipulators to grab a quantum wormhole and enlarge it to macroscopic size.

Quantum mechanics also sheds light, or possibly dark, on the paradoxes of time travel. Quantum mechanics is indeterminate – many events, such as the decay of a radioactive atom, are random. One way to make this indeterminacy mathematically respectable is the 'many worlds' interpretation of Hugh Everett III. This view of the universe is very familiar to readers of SF: our world is just one of an infinite family of 'parallel worlds' in which every combination of possibilities occurs. This is a dramatic way to describe quantum superposition of states, in which an electron spin can be both up and down at the same time, and (allegedly) a cat can be both alive and dead.*

---

* It's OK for electrons and probably nonsense for cats. See Greebo's cameo appearance in *The Science of Discworld*.

In 1991 David Deutsch argued that, thanks to the many worlds interpretation, quantum mechanical time travel poses no obstacles to free will. The grandfather paradox ceases to be paradoxical, because grandad will be (or will have been) killed in a parallel world, not in the original one.

We find this a bit of a cheat. Yes, it resolves the paradox, but only by showing that it wasn't really time travel at all. It was travel to a parallel world. Fun, but not the same. We also agree with a number of physicists, among them Roger Penrose, who accept that the 'many worlds' interpretation of quantum theory is an effective mathematical description, but deny that the parallel worlds involved are in any sense *real*. Here's an analogy. Using a mathematical technique called Fourier analysis you can resolve any periodic sound, such as the note played by a clarinet, into a superposition of 'pure' sounds that involve only one vibrational frequency. In a sense, the pure sounds form a serious of 'parallel notes', which together create the real note. But you don't find anyone asserting that there must therefore exist a corresponding set of parallel clarinets, each producing one of the pure notes. The mathematical decomposition need not have a literal physical analogue.

What about paradoxes of genuine time travel, no faffing about with parallel worlds? In the relativistic setting, which is where such questions most naturally arise, there is an interesting resolution. If you set up a situation with paradoxical possibilities, it automatically leads to consistent behaviour.

A typical thought-experiment here is to send a billiard ball through a wormhole, so that it emerges in its own past. With care, you can send it in so that when it comes (came) out it bashes into its earlier incarnation, deflecting it so that it never enters the wormhole in the first place. This is the grandfather paradox in less violent form. The question for a physicist is: can you actually set such paradoxical states up? You have to do so before the time machine is built, then build it, and see what physical behaviour actually occurs.

It turns out that, at least in the simplest mathematical formulation of this question, the usual physical laws select a unique, logically consistent behaviour. You can't suddenly plonk a billiard ball down inside a pre-existing system – that act involves human intervention, 'free will', and its relation to the laws of physics is moot. If you leave it up to the billiard ball, it follows a path that does not introduce logical inconsistencies. It is not yet known whether similar results hold in more general circumstances, but they may well do.

This is all very well, but it does beg the 'free will' question. It's a deterministic explanation, valid for idealised physical systems like billiard balls. Now, it is possible that the human mind is actually a deterministic system (ignoring quantum effects to keep the discussion within bounds). What we like to think of as making a free choice may actually be what it *feels like* when a deterministic brain works its way towards the only decision that it can actually reach. Free will may be the 'quale' of decision-making – the vivid *feeling* we get, like the vivid sense of colour we get when we look at a red flower.* Physics does not yet explain how these feelings arise. So it is usual to dismiss effects of free will when discussing possible temporal paradoxes.

This sounds reasonable, but there's a catch. The whole discussion of time machines, in physics terms, is about the possibility of people constructing the various warped spacetimes that are involved. 'Get a black hole, join it to a white one . . .' Specifically, it is about people *choosing* or *deciding* to construct such a device. In a deterministic world, either they are bound to construct it from the beginning, in which case 'construct' isn't a very appropriate word, or the thing just puts itself together, and you find out what sort of

---

* See Ian Stewart, Jack Cohen, *Figments of Reality: the origins of the curious mind* (Cambridge University Press, 1997).

universe you are in. It's just like Gödel's rotating universe: either you're in it, or you're not, and you don't get to *change* anything. You can't bring a time machine into being unless it was already implicit in the unfolding of that universe anyway.

The standard physics viewpoint really only makes sense in a world where people have free will and can choose to build, or not to build, as they see fit. So physics, not for the first time, has adopted inconsistent viewpoints for different aspects of the same question, and has got its philosophical knickers in a twist as a result.

For all the clever theorising, the dreadful truth is that we do not yet have the foggiest idea how to make a practical time machine. The clumsy and energy-wasteful devices of real physics are a pale shadow of the elegant machine of Wells's Time Traveller, whose prototype was described as 'a glittering metallic framework, scarcely larger than a small clock, very delicately made. There was ivory in it, and some transparent crystalline substance.'

There's still some R&D needed.

Probably this is a Good Thing.

NINE

AVOIDING MADEIRA

THE JOINER WAS AMAZED, AS he told his mates in the pub after work

'– so I was just finishing, and this feller comes down the ladder and says beggin' your pardon, sir, but I'd just like to check that bulkhead, please. Nothing wrong with it, says I, it's as sound as a bell. Then he says, right, right, of course, but I've just got to check something. He pulls this piece of paper out of his pocket and reads it careful, and says he's got to check that the new timber hasn't got a rare tropical worm that'll leave it looking like good wood but weaken it so much that the ship will take in too much water and will have to put in to Madeira for repairs, or something, possibly. I'll soon see about that, says I and whacks it with my hammer and, blow me, it cracks in half. I'd have sworn it was prime timber, too. Little worms everywhere!'

'Funny you should say that,' said the man opposite. 'One of 'em came up when I was working and asked if he could look at the copper nails I was usin'. Well, he takes out a knife, scrapes away at one, it's a bit of rubbish iron under a skin 'o copper! Had to do half a day's work again! Beats me how he knew. Tom said the chandler swore they were all copper when he supplied 'em.'

'Hah,' said a third man, 'one came up to me and said what would I do if a giant squid pulled the ship under. I told him I'd do nothing, being as I live in Portsmouth.' He drained his mug. 'Damned thorough, these inspectors.'

'Yeah,' said the first man, reflectively. 'They think of everything . . .'

'A goose is an inconvenient bird, I've always thought,' said Mustrum Ridcully, carving it. 'Just a bit too much for one but not quite enough for two.' He extended a fork. 'Anyone else want some? Rincewind, just get the man to send up some more oysters, will you? What do you say, gentlemen? Another six dozen? Let's push the boat out, eh? Hahah . . .'

The wizards had taken rooms at an inn, and the owner, watching the bustling staff down in his kitchen, was already thinking happily of an early retirement.

Money had not been a problem. Hex had merely teleported some from a distant bank. The wizards had debated the moral implications of this for some time, with their mouths full, but had come down in favour of the idea. They were, after all, Doing The Right Thing.

Only Ponder wasn't eating much. He nibbled a biscuit and updated his notes, before announcing: 'We have covered everything, Archchancellor. The nails, the leaking water barrels, the defective compass, the bad meat . . . there were nine reasons why the *Beagle* would have called in at the island of Madeira. Hex believes the giant squid may be a red herring. As for the nine . . . yes, I think we have assured that they will no longer occur.'

'Remind me why that's important, will you?' said the Dean. 'And pass the wine, Mustrum.'

'Without this intervention it's more than likely that Darwin will leave the ship at Madeira, should the *Beagle* call there,' said Ponder. 'He will be terribly seasick on the voyage.'

'Madeira being – ?' said the Dean.

'One of a group of islands on the way, Dean. After that it's a long haul down to the South Atlantic, round the bottom of South America with a few stops, and straight up to the Galapagos Islands.'

'Down, bottom, up,' muttered the Dean. 'How can anyone get the hang of globular navigation?'

'The phenomenon we call The Love Of Iron, sir,' said Ponder, smoothly. 'We only find it in rare metals that drop from the sky, but it's very common here. Iron here tries to point north.'

Silence fell around the table.

'North? Is that the bit at the top?' said Ridcully.

'Conventionally, sir, yes,' said Ponder, and rather foolishly added, 'but on a globe it doesn't really matter, of course.'

'Ye gods,' muttered the Dean, putting his hand over his eyes.

'How does the iron *know* which way to point?' Ridcully persisted. 'Metal can't think.'

'It's a bit like . . . like peas turning to follow the Sun, sir,' hazarded Ponder, not sure if they actually did; perhaps it was pea farmers.

'Yes, but peas are living things,' said Ridcully. 'They . . . know about the Sun, right?'

'Peas aren't exactly renowned for their brains, Archchancellor,' said the Chair of Indefinite Studies, 'hence the term pea-brained.'

'But a pea must be a bloody genius compared to a lump of iron, yes?' said Ridcully.

Ponder knew he had to put a stop to this. The wizards were still determined to apply common sense to Roundworld, and that would get them nowhere.

'It's a force that can occur on globe-shaped worlds,' he said. 'It's caused by the molten iron core spinning, and helps prevent life on the surface being fried by the Sun.'

'Sounds like Deitium in disguise, doesn't it,' said Ridcully. 'Planet gets this big magical umbrella so that life can survive? Shows forethought.'

'It doesn't work quite like that, Archchancellor,' said Ponder. 'Life evolved because conditions allowed it to do so.'

'Ah, but if conditions hadn't been right, there wouldn't have been any life,' said Ridcully. 'Therefore the whole exercise would have been pointless.'

'Not really, sir. There wouldn't have been anyone to point out the pointlessness of it,' said Ponder. 'I was about to add that some birds, like pigeons, use The Love Of Iron to help them navigate long distances. They have tiny things called "magnets" in their head, says Hex. They're . . . little bits of iron that know where the North Pole is . . .'

'Ah, I know that bit,' said the Lecturer in Recent Runes. 'The North and South Poles are those bits on a globe where the spindle comes out. But they're invisible, of course,' he added.

'Um,' said Ponder.

'Just a minute, can we get back to these birds?' said Ridcully. 'Birds with magnet heads?'

'Yes?' said Ponder, knowing that this was going to be loaded.

'How?' said Ridcully, flourishing a goose leg. 'On this globe, birds grew out of great big monstrous lizard beasts, isn't that so?'

'Er . . . *small* great monstrous beasts, sir,' said Ponder, wishing not for the first time that his Archchancellor did not have a knack for remembering inconvenient details.

'Did *they* have to fly long distances through fog and bad weather?' said the Archchancellor.

'I doubt it, sir,' said Ponder.

'So did they already have these magnets in their heads from day one, or did they turn up by some godly hand? What does Mr Darwin of *The Origin* say about that?'

'Not very much, sir,' said Ponder. It had been a long day.

'But it suggests, does it not, that *The Ology* haha, is right and *The Origin* is wrong. Perhaps the magnets were added when needed?'

'Could be, sir,' said Ponder. Just don't let him start on the eyeball, he thought.

'I've got a question,' said Rincewind, from the end of the table.

'Yes?' said Ponder, quickly.

'There's going to be monster creatures on these island we're heading for, yes?'

'How did you know that?' said Ponder.

'It just came to me,' said Rincewind gloomily. 'So there *are* monsters?'

'Oh, yes. Giants of their kind.'

'With big teeth?'

'No, not really. They're tortoises.'

'How big?'

'About the size of an easy chair, I think.'

Rincewind looked suspicious.

'How fast?'

'I don't know. Not very fast.'

'And that's *it*?'

'From a Darwinian perspective, the islands are famous for their many species of finches.'

'Any of them carnivorous?'

'They eat seeds.'

'So . . . there's *nothing* dangerous where we're going?'

'No. Anyway, we don't have to go there. All we have to do now is find the point where he decides to write *The Ology* instead of *The Origin*.'

Rincewind pulled the dish of potatoes towards him.

'Sez you,' he said.

+++ I need to communicate grave news +++

The words came out of the air. In Roundworld, Hex had a voice.

'We're having a bit of a celebration here,' said Ridcully. 'I'm sure your news can wait, Mr Hex!'

+++ Yes. It can +++

'Good. In that case, Dean would you pass me—'

+++ I would not wish to spoil your appetite +++ Hex went on.

'Glad to hear it.'

+++ The destruction of the human race can wait until after the pudding +++.

Ridcully's fork hovered between his plate and his mouth. Then he said: 'Would you care to explain this, please, Mr Stibbons?'

'I can't, sir. What is happening, Hex? We completed all those tasks properly, didn't we?'

+++ Yes. But, pause for significance, have you heard of a mythical creature called a, pause again, hydra? +++

'The monster with many heads?' said Ponder. 'You don't need to tell us when you pause, by the way.'

+++ Thank you. Yes. Cut off one head and a dozen grow in its place. This history is a hydra +++

Rincewind nodded at Ponder. 'Told you,' he said, with his mouth full.

+++ I am unable to explain why this is the case, but there are now 1457 reasons why Darwin did not write *The Origin of Species*. The book has never been written in this history. The voyage has never taken place +++

'Don't be silly! We know it did!' said the Dean.

+++ Yes. It did. But now, it hasn't. Charles Darwin the scientist has been removed from this history while you ate. He was, and now he was not. He became a little-remembered priest who caught butterflies. He wrote no book. The human race dies in five hundred years +++

'But yesterday—' Ridcully began.

+++ Consider time not as a continuous process but as a succession of discrete events. Darwin's scientific career has been excised. You remember him, but that is because you are not part of this universe. To deny this is simply to scream at the monkeys in the next tree +++

'Who did it?' said Rincewind.

'What sort of question is that?' said Ponder. 'No one *did* it. There isn't anyone to *do* things. This is some kind of strange phenomenon.'

+++ No. The act shows intelligence +++ said Hex. +++ Remember, I detected malignity. I surmise that your interference in this history has led to some counter-measure +++

'Elves again?' said Ridcully.

+++ No. They are not clever enough. I can detect nothing except natural forces +++

'Natural forces aren't animate,' said Ponder. 'They can't think!'

+++ pause for dramatic effect . . . Perhaps the ones here have learned to +++ said Hex.

TEN

WATCH-22

 IN THE STANDARD VERSION OF Roundworld history, Charles Darwin's presence on the *Beagle* came about only because of a highly improbable series of coincidences – so improbable that it is tempting to view them as wizardly intervention. What Darwin expected to become was not a globetrotting naturalist who revolutionised humanity's view of living creatures, but a country vicar.

And it was all Paley's fault.

Natural Theology's seductive and beautifully argued line of reasoning found considerable favour with the devout people of Georgian (III and IV) England, and after them, the equally devout subjects of William IV and Victoria. By the time Victoria ascended to the throne, in 1837, it was indeed almost compulsory for country vicars to become experts in some local moth, or bird, or flower, and the Church actively encouraged such activities because they were continuing revelations of the glory of God. The Suffolk rector William Kirby was co-author, with the businessman William Spence, of a lavish four-volume treatise *An Introduction to Entomology*, for example. It was fine for a clergyman to interest himself in beetles. Or geology, a relatively new branch of science that had grabbed the young Charles Darwin's attention.

The big breakthrough in geology, which turned it into a fully fledged science, was Charles Lyell's discovery of Deep Time – the idea that the Earth is enormously older than Ussher's 6000 years. Lyell argued that the rocks that we find at the Earth's surface are the product of an ongoing sequence of physical, chemical, and biological processes. By measuring the thickness of the rock layers, and estimating the rate at which those layers can form, he deduced that the Earth must be extraordinarily ancient.

Darwin had a passion for geology, and absorbed Lyell's ideas like a sponge. However, Charles was basically rather lazy, and his father knew it. He also knew, to quote Adrian Desmond and James Moore's biography *Darwin*, that:

> The Anglican Church, fat, complacent, and corrupt, lived luxuriously on tithes and endowments, as it had for a century. Desirable parishes were routinely auctioned to the highest bidder. A fine rural 'living' with a commodious rectory, a few acres to rent or farm, and perhaps a tithe barn to hold the local levy worth hundreds of pounds a year, could easily be bought as an investment by a gentleman of Dr. Darwin's means and held for his son.

That, at least, was the plan.

And at first, the plan seemed to be working. In 1828 Charles was admitted to the University of Cambridge, taking his oath of matriculation one cold January morning, swearing to uphold the university's ancient statutes and customs, 'so help me God and his holy Gospels'. He was enrolled at Christ's College for a degree in theology, alongside his cousin William Darwin Fox who had started the previous year. (Charles had previously attempted medicine in Edinburgh, following in the footsteps of his father and grandfather, but he became disillusioned and left without a degree.) After getting his Batchelor of Arts degree, he might spend a further year reading theology, ready to be ordained in the Anglican Church. He could become a curate, marry, and take up a rural position near Shrewsbury.

It was all arranged.

Shortly after starting at Christ's, Charles was bitten by the beetle bug, as it were. *An Introduction to Entomology* sparked off an intense interest in beetles, when seemingly half the nation was out searching the woods and hedgerows to find new species. Since there were more species of beetle in the world than anything else, this was a serious prospect. Charles and his cousin scoured the byways of rural Cambridgeshire, pinning their catches in neat rows on large sheets of cardboard. He didn't find a new species of beetle, but he found a rare German one, seen only twice before in the whole of England.

Towards the end of his second year at university, exams loomed. Darwin had been too intent on beetles and a young lady named Fanny Owen and had neglected his academic studies. Now he had a mere two months left to do the work of two years. In particular, there would be ten questions on the book *Evidences of Christianity*, by one William Paley. Darwin had already read the book, but now he read it again with new attention – and loved it. He found the logic fascinating. Moreover, Paley's political leanings were distinctly left-wing, which appealed to Charles's innate sense of social justice. Bolstered by his studies of Paley, Darwin scraped through.

Next in line were the final exams. Another of Paley's books was on the syllabus: *Principles of Moral and Political Philosophy*. The book was outdated, and sailed close to the wind of (political) heresy and well into the shallows of unorthodoxy; that was *why* it was on the syllabus. You had to be able to argue the case against it, where applicable. It said, for example, that an established Church formed no part of Christianity. Darwin, then a very conventional Christian, wasn't sure what to think. He needed to broaden his reading, and in so doing he selected yet another book by his idol Paley: *Natural Theology*. He knew that many intellectuals derided Paley's stance on design as naive. He knew that his own grandfather, Erasmus Darwin, had held a radically different view, speculating about spontaneous changes in organisms in his own book *Zoonomia*. Darwin's sympathies were with Paley, but he started wondering how scientific laws were established, and what kind of evidence was acceptable, a quest that led him to a book by Sir John Herschel with the mind-numbing title *Preliminary Discourse on the Study of Natural Philosophy*. He also picked up a copy of Alexander von Humboldt's *Personal Narrative*, a 3754-page blockbuster about the intrepid explorer's trip to South America.

Darwin was entranced. Herschel stimulated his interest in science, and Humboldt showed him how exciting scientific discoveries could be. He determined, then and there, to visit the volcanoes of the Canary Islands and see for himself the Great Dragon Tree. His friend Marmaduke Ramsay agreed to accompany him. They would leave for the tropics once Darwin had signed the 39 Articles of the Anglican Church at his degree ceremony. To prepare for the journey, Charles went to Wales to carry out geological fieldwork. He discovered that there was no Old Red Sandstone in the Vale of Clwyd, contrary to the current national geological map. He had won his geologists' spurs.

Then a message arrived. Ramsay had died. The Canary scheme shuddered to a halt. The tropics seemed further away than ever. Could Charles go it alone? He was still trying to decide when a bulky package arrived from London. Inside was a letter, offering him the opportunity to join a voyage round the world. The ship would sail in a month.

The British Navy was planning to explore and map the coast of South America. It was to be a chronometric survey, meaning that all navigation would be done using the relatively new and not fully trusted technique of finding longitude with the aid of a very accurate watch or chronometer. A 26-year-old sea captain, Robert FitzRoy, would head the expedition; his ship would be the *Beagle*.

FitzRoy was worried that the solitude of his command might drive him to suicide. The risk was not far-fetched: the *Beagle*'s former captain Pringle Stokes had shot himself while mapping a particularly convoluted bit of the coast of South America. Further, one of FitzRoy's uncles had slit his own throat in a fit of depression.* So he had decided that he needed someone to talk to, to keep him sane. It was this position that was now being offered to Darwin. The job would be especially suitable for someone with an interest in natural history, and the ship had the necessary scientific equipment. Technically, Darwin would not be 'ship's naturalist', as later he sometimes claimed, and that presumption would eventually lead to an almighty row with the *Beagle*'s surgeon Robert McCormick, because by tradition the surgeon did the job of naturalist in his spare time. Darwin was being hired as a 'gentleman companion' for the captain.

Charles decided to accept the offer, but his father, forewarned by Charles's sisters, refused permission. Darwin could have gone against his father's wishes, but the thought made him feel very uncomfortable, so he wrote to the Navy and turned the job down. Then, uncharacteristically, his father opened a loophole – our first example of what looks suspiciously like wizardly interference. Charles might yet be allowed to go, he said, provided 'some person of good standing' recommended it. Both Charles and his father knew who was meant: Uncle Jos (Wedgwood, grandson of the founder of the pottery company). Jos was an industrialist, and Dr Darwin trusted his judgement. So Charles and his uncle sat up very late, composing a suitable letter. Jos told Dr. Darwin that such a voyage would be the making of the young man. And, slyly, he added that it would improve his knowledge of natural history, which would be very useful for a subsequent career in the clergy.

Darwin Senior relented (score one to the wizards). Excited beyond measure, Charles hurriedly wrote another letter to the Navy, this time accepting. But then he heard from FitzRoy, who told him that the post was no longer vacant. The captain had given it to a friend.

---

* In 1865 FitzRoy did exactly the same, having been turned down for a promotion. Narrativium at work?

However, Darwin was top of the list if his friend changed his mind.

Darwin went to London, to make contingency plans in case he got lucky, and to keep an appointment with FitzRoy. He arrived to be told that the captain's friend had changed his mind, not five minutes earlier. (Wizards again?) His wife had objected to the length of the voyage, then planned to be three years. Did Darwin still want the job?

Lost for words, Charles nodded.

Darwin's heart sank when he saw the ship. The *Beagle* was a rotting, eleven-year-old brig, with ten guns. It was being rebuilt, partly at FitzRoy's own expense, so it would be seaworthy enough. But the ship was cramped, a mere 90 feet (30m) long by 24 feet (8m) wide. Could his companionship with the captain survive such a lengthy voyage in such close contact? Fortunately, he was allocated one of the larger cabins.

The *Beagle*'s assignment was to survey the southern end of South America, in particular the complicated islands around Tierra del Fuego. The Admiralty had provided 11 chronometers for navigation, because the trip would be the first attempt to circumnavigate the Globe using marine chronometers to find longitude. FitzRoy borrowed five more, then bought six himself. So the *Beagle* sailed with a massive 22 chronometers on board.

The voyage started badly. Darwin was sick as a dog, crossing the Bay of Biscay, and had to endure the sound of sailors being flogged as he lay nauseated in his hammock. FitzRoy was hot on discipline, especially at the beginning of a voyage. Privately, the captain expected his 'companion' to jump ship the moment it touched land, and hotfoot it back to England. The ship was supposed to put in at Madeira to take on fresh food, which would be the perfect opportunity. But the Madeira landing was cancelled because the sea was too heavy and there was no pressing need (score 3 to the wizards?).

Instead, the *Beagle* headed for Tenerife in the Canaries. If Charles jumped ship there, he could see the volcanoes and the Great Dragon Tree. But the consul in Santa Cruz was scared that visitors from England might introduce cholera to his islands, and he refused the *Beagle* permission to put into port without undergoing quarantine (score 4? We'll see). Unwilling to wait off land for the required two weeks, FitzRoy ordered the *Beagle* south, to the Cape Verde Islands.

It may not have been the wizards at work, but *something* was determined that Charles should stay on the *Beagle*. And now, a fifth coincidence, involving his great love, geology, made it impossible for him to do anything else. As the *Beagle* sailed westward, the ocean grew calm, the air warm. Darwin could trawl for plankton and jelly-fish with home-made gauze nets. Things were looking up. And when they finally touched land, the island of St Jago in the Cape Verde Islands, Darwin found it hard to believe his luck. St Jago was a rugged volcanic outcrop, with conical volcanoes and lush valleys. Charles could do geology. And natural history.

He collected everything. He noticed that an octopus can change colour, and mistakenly thought this was a new discovery. After two days, he had worked out the geological history of the island, using the principles he had learned from Lyell. Lava had flowed over the seabed, trapping shells and other debris, and had later been raised to the surface. All of this must have happened relatively recently, because the shells were just like the fresh ones lying on the beach. This was not the conventional theory of the day, which held that volcanic structures were incredibly old.

The young man was coming into his own.

In the end, the voyage lasted five years, and in the whole of that time, poor Darwin never found his sea-legs. Even on the final run home, he was still seasick. But he contrived to spend most of the voyage on land, and only 18 months at sea. And while on land, he made discovery after discovery. He found fifteen new species of flat-worm in Brazil. He studied rheas, giant flightless birds related to the ostrich, in Argentina. There, too, he found fossils, including the head of a giant armadillo-like glyptodont. In Tierra del Fuego he turned anthropologist, and studied the people. 'I shall never forget how savage & wild one group was,' he wrote, on encountering 'naked savages'. He found more fossils, among them bones of the ground-sloth *Megatherium* and the llama-like *Macrauchenia*. In Chile, he studied the geology of the Andes and decided that they, and the plains beyond, had been thrust skyward in some gigantic geological upheaval.

From the South American mainland, the *Beagle* went north-west to the Galápagos, a tight group of a dozen or so islands, far out into the Pacific ocean. The islands had fascinating geology, mainly vol-canic, and a great variety of animals that were not found anywhere else. There were the spectacular giant tortoises that had given the islands their name. Darwin measured the circumference of one as seven feet (2m). There were iguanas, and birds – boobies, warblers, finches. The finches had beaks of different shapes and sizes, depend-ing on the food they ate, and Darwin divided them up into a series of subfamilies. He did not notice that different types of animals occurred on different islands, until Nicholas Lawson pointed this out. (The wizards again? Oh yes, this will have happened soon . . .) But he did notice that the mockingbirds of Charles and Chatham islands (now Santa Maria and San Cristobál) were different species, and when, now alerted, he looked on James Island (San Salvador), he found yet a third species. But Darwin was not greatly interested in small variations in species, or how those variations corresponded to the local geography. He was vaguely aware of some theorising about species change, or 'transmutation', if only from his grandfather Erasmus, but the topic didn't interest him and he saw no reason to collect evidence for or against it.

And so the *Beagle* continued to Tahiti, New Zealand, and Australia.

Darwin had seen wonders that would shortly revolutionise the world. But he did not yet understand what he had seen.

In Tahiti, though, he glimpsed his first coral reef. Before leaving Australia, he was determined to find out how coral islands came into being. Lyell had suggested that because the coral animals live only in shallow waters, with ample sunlight, the reefs must be built on top of submerged volcanoes. This also explained their ringed shape. Darwin didn't believe Lyell's theory. 'The idea of a lagoon island, 30 miles in diameter being based on a submarine crater of equal dimensions, has always appeared to me to be a monstrous hypothesis.' Instead, he had his own theory. He already knew that land could rise, he'd seen that in the Andes. He reasoned that if some land went up, then other land ought to go down, to maintain the balance of the Earth's crust. Suppose that when the reef started to form, the water was shallow, but then the ocean floor started descending slowly, while the coral polyps at the surface continued building the reef. Then eventually you would get a huge mountain of coral rising from what was now the ocean depths – all built by tiny creatures, always in shallow water while the building was going on. The shape? That was the result of an island with a fringing reef collapsing. The island would sink, leaving a hole in the middle, but the reef would continue to grow.

Five years and three days after the *Beagle* set sail from Plymouth, Darwin walked into the family home. His father glanced up from his breakfast. 'Why,' he said, 'the shape of his head is quite altered.'

Darwin did not come up with the concept of evolution during his *Beagle* voyage. He was too busy amassing specimens, mapping geology, taking notes, and being seasick, to have time to organise his observations into a coherent theory. But when the voyage was over, he was promptly elected to the Royal Geological Society. In January 1837 he presented his inaugural paper, on the geology of Chile's coast. He suggested that the Andes mountains had originally been the ocean floor, but had later been uplifted. His diary records amazement at 'the wonderful force which has upheaved these mountains, & even more so the countless ages [needed] to have broken through, removed & levelled whole masses of them'. Much later, the Chilean coast became part of the evidence for 'continental drift': we now think that these mountains result from subduction, as the Nazca tectonic plate slides underneath the South American plate.

Darwin could certainly spot them.

His interest in geology had other, less obvious, implications. He was starting to wonder about the finches of the Galápagos. They seemed to contradict Lyell's view that local geological conditions determined what species were created. It was a puzzle.

In fact, it was more of a puzzle than Darwin thought, because he had misunderstood the finches completely. He thought they all fed on the same food, in big flocks. He had not noticed important differences among their beaks, and he even had trouble identifying different species. Some, he believed, were not finches at all, but wrens and blackbirds. He was so baffled by the birds, and so indifferent to the specimens he had collected, that he donated the lot to the Zoological Society. Within ten days the Society's bird expert John Gould had worked out that they were all finches, all very closely related, forming a tightly knit grouping that nonetheless contained twelve* distinct species. This number was surprisingly large for such a small group of tiny islands. What had caused such diversity? Gould wanted to know, but Darwin didn't care.

By 1837, Paley's logic was no longer in vogue. The scientifically literate theist now believed that God had set up the laws of nature at the time of Creation, and that those laws included not just the 'background' laws of physics, to which Paley subscribed, but also

---

* Now considered to be thirteen, plus a fourteenth on the Cocos Islands. (Look, people write and *complain* if we don't point this kind of thing out.)

the development of living creatures, which Paley had denied. The laws of the universe were fixed for all eternity. They had to be, otherwise God's creation was flawed. Paley's analogies were used against him. What kind of artificer made such bad machinery that He had to keep tinkering with it all the time to keep it working?

Science and theology were ripping asunder. The political corruption of the Church was becoming undeniable; now its intellectual claims were also coming under fire. And some radical thinkers, often medics who had studied comparative anatomy and noticed remarkable similarities between the bones of entirely different animals, were engaged in speculation that changed the view of creation itself. According to the Bible, God had created each type of animal as a one-off item – whales and winged fowl on the fifth day, cattle and creeping things and humans on the sixth. But these medical types were starting to think that species could change, 'transmute'. Species were not fixed for all time. They realised that there was a rather big gap between, say, a banana and a fish. You couldn't cross that gap in one step. But given enough time, and enough steps . . .

Darwin slowly became caught up in the flow. His Red Notebook, where he recorded anything that he saw or that came to mind, began to hint at the 'mutability of species'. The hints were incomplete and ill-assorted. Deformed babies resembled new species. The beaks of Galápagos finches were of different shapes and sizes. Rheas were a puzzle, though: two distinct species of the giant birds had overlapping ranges in Patagonia. Why didn't they merge into a single species?

By July, he had secretly started a new notebook, his B Notebook. It was on the transmutation of species.

By 1839 Darwin was building up a complete picture, and he wrote a 35-page summary of his thinking. A crucial influence was Thomas Malthus, whose 1826 *Essay on the Principle of Population* pointed out that the unchecked growth of organisms is exponential (or 'geometric', in the old-fashioned phrase of the time), whereas that of resources is linear ('arithmetic'). Exponential growth occurs when each step multiplies the size by some fixed amount, for example 1, 2, 4, 8, 16, 32, where each number is twice the previous one. Linear growth adds some fixed amount at each step, for instance 2, 4, 6, 8, 10, where each number exceeds the previous one by 2. However small the multiplier is in exponential growth, provided it is bigger than 1, and however large the number added in linear growth may be, it turns out that in the long run exponential growth *always* beats linear. Though it does take some time if the multiplier is close to 1 and the number being added is huge.

Darwin had taken on board Malthus's argument, and he had realised that in practice what keeps populations down is competition for resources, such as food and a place to live. This competition, he wrote, leads to 'natural selection', in which those creatures that are victorious in the 'war of nature' are the ones that produce the next generation. Individual creatures within a species are not exactly identical; those differences make it possible for the force of natural selection to produce slow, gradual changes. How far might such changes go? In Darwin's view, very far indeed. Far enough to lead to entirely new species, given enough time. And thanks to geology, scientists now knew that the Earth was very, very old.

Darwin, following family tradition, was a Unitarian. This particular branch of Christianity has been aptly described as 'people who believe in at most one God'. As a sound Unitarian, he believed that the Deity must work on the grandest of scales. So he finished his summary with a powerful appeal to the Unitarian view of the Deity:

> It is derogatory that the Creator of countless systems of worlds should have created each of the myriads of creeping parasites and slimy worms which have swarmed each day of life on land and water on this one globe. We cease being astonished, however much we may deplore, that a group of animals should have been directly created to lay their eggs in bowels and flesh of others – that some organisms

should delight in cruelty . . . From death, famine, rapine, and the concealed war of nature we can see that the highest good, which we can
conceive, the creation of the higher animals has directly come.

God surely has better taste than to create nasty parasites *directly*.
They exist only because they are a necessary step along the path that
leads to cats, dogs, and us.

Darwin had his hypothesis.

Now he began to agonise about how to bring it to the waiting
world.

ELEVEN

WIZARDS ON
THE WARPATH

IN THE GLOOM OF THE High Energy Magic building, Hex wrote. Every minute another page slid off the writing table.

'"Boat sunk by collision with Spanish fishing vessel",' Ponder Stibbons read out, a tremor in his voice. '"Boat shipwrecked on uncharted reef near Madeira. Boat found drifting minus all crew, with the table laid for a meal. Boat catches fire, all lost. Boat struck by meteorite. Darwin accidentally shot by ship's surgeon and naturalist during a collecting expedition on the island of St Jago. Darwin accidentally shot by ship's captain. Darwin accidentally shot by himself. Darwin loses place on boat. Darwin leaves boat because of seasickness. Darwin loses notebooks. Darwin stung to death by wasps! Darwin bangs head on underside of table and loses memory . . ."' He put down the paper. 'And these are the more sensible causes.'

'The stone dropping out of the sky was sensible?' said Ridcully.

'Compared to the attack of the giant squid, Archchancellor, I would say so,' said Ponder. 'And the enormous waterspout. And the shipwreck off the coast of Norway.'

'Well, ships do get wrecked,' said the Dean.

'Yes, sir. But the country known as Norway is in the wrong direction. The *Beagle* would only get there by sailing backwards. Hex is right, sir. This is insane. The moment that we decided to change one simple little history, the whole of the universe is trying to stop the voyage happening! And mathematically speaking, this is illegal!'

Ponder thumped the table, his face red. The senior wizards shied. This was as unnerving as hearing a sheep roar.

'My word!' said Ridcully. 'Is it?'

'Yes! There *must* be room in phase space for the possibility that *Origin* gets written! It's not against the physical laws of this universe!'

'That a young inexperienced man takes a voyage around this world and has an insight that changes mankind's view of itself?' said the Dean. 'You must admit it looks a bit unlikely – sorry, sorry, sorry!' He backed away as Ponder advanced.

'One of the biggest religions on Roundworld was founded by a carpenter's son!' Ponder snarled. 'For *years*, the most powerful person on the planet was an actor! There's got to be room for Darwin!'

He stamped back to the table and picked up a handful of papers. 'Look at this stuff! "Darwin bitten by poisonous spider . . . Darwin savaged by kangaroo . . . stung by jellyfish . . . eaten by shark . . . *Beagle* found floating, table laid for a meal, this time in a different ocean, still no one on board . . . Darwin struck by lightning . . . killed by volcanic activity . . . *Beagle* sunk by freak wave" . . . *does anyone expect us to believe this for one minute?*'

There was a ringing silence.

'I can see this is worrying you, Mr Stibbons,' said Ridcully.

'Well, yes, I mean, *yes*, it's so . . . wrong! The multiverse is not supposed to change the rules. Anything that's possible to happen has a universe for it to happen in! I mean, *here*, yes, the rules can be bent in all kinds of ways, but in Roundworld there's no one to bend them!'

'I've got an idea,' said Rincewind. The other wizards turned, amazed at this revelation.

'Yes?' said Ponder.

'Why not just take it for *granted* that someone is out to get you?' said Rincewind. 'That's what I do. Don't bother to work out the fine detail. Look, when you first started to tinker, it was all going to be plain sailing, right? Make a few little adjustments, pinch a fish, and it'd all be OK? But now there are nearly fifteen hundred new reasons—'

With a rattle, Hex's writing desk started up. The pens wrote:

+++ 3563 reasons now +++

'They're breeding!' said Ridcully.

'There you are, then,' said Rincewind, almost cheerfully. 'Something down there is frightened. It's so frightened that it's not even going to let him get on the boat. I mean, he has to take the voyage *whatever* book he writes, isn't that right!'

'Yes, of course,' said Ponder. *Theology of Species* gets taken seriously because it's written by a renowned and respected scientist whose research was meticulous. So was *The Origin*. Either way, he needs to be on that boat. But the moment we take an interest, the voyage doesn't happen!'

'Then if it was me, I'd say that something's got really worried,' said Rincewind. 'They don't mind if *The Ology* doesn't get written in just one universe, but they hate the idea of *The Origin* being written at *all*.'

'Oh, *really*?' said Ridcully. 'The nerve! I am the master of this college, and that – ' he pointed to the little globe ' – is university property! Now I'm getting angry. We're going to fight back, Mr Stibbons!'

'I don't think you can fight a whole universe, sir!'

'It's the prerogative of every life form, Mr Stibbons!'

Gales roared for three weeks. Roundworld time was mutable for the wizards; it only affected them if they wanted it to.

Something or someone didn't want the *Beagle* to sail, and they could influence the weather. They could influence anything. But of them, there was still no sign.

The Dean watched the storm in the big omniscope in the HEM.

'That's what happened when Darwin gets on board in this universe,' said Ponder, adjusting the omniscope. 'If he hadn't gone, his place is taken by an artist, who produced a famous portfolio as a result. His name was Preserved J. Nightingale. You met his wife.'

'Preserved?' said the Dean, watching the dismal gale.

'Short for Preserved-by-God,' said Ponder. 'He was found as a child in the wreckage of a ship. His adopted parents were very religious. And . . . ah yes . . . *this* is the weather they get when *he* is on board.'

The omniscope flickered.

'No gale?' said the Dean, looking at the blue sky.

'Brisk winds from the north-east. They're ball-world directions, sir. For the purposes of the voyage, they are ideal. I see you have your "Born to Rune" jacket on, sir.'

'We've got a fight on our hands, Stibbons,' said the Dean, severely. 'It's a long time since I've seen the Archchancellor so angry at anyone but me! Have you finished?'

'Just finishing, sir,' said Ponder.

The HEM had a deserted look. That was because it had been, by and large, deserted. Thick tubes led out from Hex, across the floor and out over the lawn towards UU's Great Hall.

The wizards were going to war. It took a lot to make that happen, but you couldn't let any old universe push you around. Gods, demons and Death were one thing, but mindless matter shouldn't be allowed to get ideas.

'Couldn't we just find a way to bring Darwin back here?' said the Dean, watching Ponder prod buttons on Hex's keyboard.

'Quite probably, sir,' said Ponder.

'Well, then, why don't we just bring him here, explain the situation, and drop him off on his island? We could even give him a copy of his book.'

Ponder shuddered.

'There are quite a lot of reasons why that course of action might not, with ease, be rescued in any coherent way from the category of the insanely unwise, Dean,' he said, having worked out that the senior wizards lost interest in any sentence that went on past twenty words. 'For one thing, he'd know.'

'We could bop him on the head,' said the Dean. 'Or put a 'fluence on him. Yes, that'd be a good idea,' he said, because it was his. 'We could sit him in a comfy chair and read out the right book to him. He'd wake up back home and think he's made it all up.'

'But he wouldn't have been there,' said Ponder. He waved a hand. In the air overhead, a little ball of multicoloured light appeared. It looked like a tangle of glowing strings, or the mating of rainbows.

'Oh, we could sort that out,' said the Dean airily. 'Stick some sand in his boots, a few finch feathers in his pocket . . . we are wizards, after all.'

'That would be unethical, Dean,' said Ridcully.

'Why? We're the Good Guys, aren't we?'

'Yes, but that rather hinges on doing certain things and not doing others, sir,' said Ponder. 'Playing around with people's heads against their will is almost certainly one of the nots. You should get ready to move quickly, sir.'

'What are *you* doing, Stibbons?'

'I've got Hex to cast a thaumatic glyph in conditional Darwin space,' said Ponder. 'But to resolve it properly Hex will have to run the thaumic reactor a little higher than usual.'

'How much higher?' said the Dean suspiciously.

'About 200 per cent, sir.'

'Is that safe?'

'Absolutely not, sir. Hex, glyphic resolution in twenty seconds. Dean, run! Run, sir!'

From the direction of the Old Squash Court came a sound that had been there all the time, unheeded, and was now growing louder. It was the *whum whum* of dying thaums, each one yielding up its intrinsic magic . . .

Wizards have a wonderful turn of speed.

Ponder and the Dean reached the Great Hall in twelve seconds, the Dean slightly in the lead. The ball of rainbows had got there before them, though, and hung high over the black and white flagstones of the floor.

The hall was packed with wizards. Teams had been sent out to the furthest corners of the university, which were pretty far. Space and time had long ago been warped by the ancient magical stones, and there were wizards at UU who had happily occupied nooks and corners for decades or longer, regarding the Great Hall and surrounding buildings as the colonists on some faraway continent might regard the ancient mother country. Distant studies had been broken into and their occupants dragged out or, in some unfortunate cases, swept up. Wizards that Ponder had never seen before were in the throng, blinking in the light of common day.

Panting slightly, Ponder hurried over to Ridcully.

'You said you wanted a map, sir,' he said.

'Yes, Stibbons. Can't plan a campaign without a map!'

'Then look up now, sir! Here it comes!'

The air wavered for a moment, and then the mated rainbows gave birth. Frozen streamers of light looped through the hazy air of the hall. They twisted and tangled and curved in ways that suggested more than the everyday four dimensions were involved.

'Looks very pretty,' said the Archchancellor, blinking. 'Er . . .'

'I thought it would help us sort out further nodalities,' said Ponder.

'Ah yes, good idea,' said Ridcully. 'No one wants unsorted nodalities.' The other senior wizards nodded sagely.

'*By which I mean,*' Ponder added, 'it will show us those points where our intervention will have been going to be was essential, if I can put it that way.'

'Oh,' said the Archchancellor. 'Er . . . what does the coloured line mean, exactly?'

'Which one, sir?'

'All of them, man!'

'Well, the points of intervention that require a human show up as red circles. Those that can be left to Hex are white. The blue lines represent the author of, ahem, *The Ology*, the yellow lines is the optimum path for the author of *The Origin*, and the green line represent slippage between futures. Known thaumic occlusions are purple, but I expect you worked that out already.'

'What's that one?' said the Dean, pointing to a red circle with his staff.

'We must make certain he doesn't get off the boat at an island called Tenerife,' said Ponder. 'Seasickness again, you see. Quite a few Darwins got off there.'

The tip of the staff moved. 'And that one?'

'He *must* get off the boat at the island of St Jago. He has valuable insights there.'

'Sees things evolvin', that kind of thing?' said Ridcully.

'No, sir. You can't see things evolving, even when they're doing it.'

'We saw them on Mono Island,' said the Lecturer in Recent Runes. 'You could practically *hear* them!'

'Yes, sir. But we have a *god* of evolution. Gods aren't patient. On Roundworld, evolution takes time. Lots of time. Darwin was raised in the belief that the Roundworld universe was created in six days –'

'Which is correct, as I have pointed out,' said the Dean proudly.

'Yes,' said Ponder, 'but I have also pointed out that on the inside it took billions of years. It is vital that Darwin realises that evolution has got lots of time to work in.'

Before the Dean could protest, Ponder turned back to the shining, twisting tangle of light.

'*There* is where the mast falls on his head in the port of Buenos Aires,' he said, pointing. 'The *Beagle* was shot at. It was meant to be a blank, fired from a cannon, but for some reason it had been loaded. The British were very upset about it, and issued a stern diplomatic protest by sending a warship to bombard the port to rubble. This one is where Darwin bludgeons himself into unconsciousness with his own bolas in Argentina. This one is where he's severely injured putting down an insurrection—'

'He got about a bit for a man who collected flowers and things,' said Ridcully, with a touch of admiration.

'Look, I've been thinking about all this,' said the Dean. 'This "science" is all about the search for truth, yes? Why don't we just tell them the truth?'

'You mean tell them that their universe was accidentally started by you, Dean, sticking your hand into some raw firmament created to use up spare power from the thaumic reactor?' said Ridcully.

'Put like that it seems a bit unlikely, I admit, but—'

'No direct contact, Dean, we agreed about that,' said Ridcully. 'We just clear his way. What's *that* nodality, Stibbons? It's flashing.'

Ponder looked at where the Archchancellor's staff was pointing.

'That's a tricky one, sir. We will have to ensure that Edward Lawson, a British official in the Galápagos Islands, isn't struck by a meteorite. It's a new malignity, Hex says. In a number of histories, it happens a few days before he meets Darwin. Remember, sir? I mentioned it in my yellow folder that was delivered to your office this morning.' Ponder sighed. 'He draws Darwin's attention to some interesting facts.'

'Ah, I read that one,' said Ridcully, his happy tone indicating that this was a lucky coincidence. 'Darwin seemed to be too busy runnin' around like a monkey in a banana plantation to spot the clues, eh?'

'It would be true to say that his full theory of natural selection was evolved on mature reflection some time after his voyage, yes,' said Ponder, carefully answering a slightly different question.

'And this chap Lawson was important?'

'Hex believes so, sir. In a way, everyone Darwin met was important. And everything he saw.'

'And then whoosh, this chap was hit by a rock? I call that suspicious.'

'Hex does too, sir.'

'I'll be jolly glad when we've got this Darwin to the damn islands, then,' said the Archchancellor. 'We'll need a holiday after this. Oh well, I'll address the wizards now. I hope we'll have enough for—'

'Er, we haven't just got to get him to the islands. We've got to get him all the way home, sir,' said Ponder. 'He'll be away from home for nearly five years.'

'Five years?' said the Dean. 'I thought visiting the wretched islands was what it was all about!'

'Yes and then again, in a very real sense, no, Dean,' said Ponder. 'It would be more correct to say they later *became* what it was all about. He was actually there for a little more than a month. It was a very long voyage, sir. They went all around the world. I'm sorry, I hadn't made that clear. Hex, show the *entire* timelines, please.'

The display began to recede, drawing from nowhere more and more tangles and loops, as if half a dozen cosmic kittens had been given stars to play with instead of balls of wool. There was a gasp from the throng of wizards.

The tangles were still streaming away overhead when the Dean said: 'There's *millions* of the wretched things!'

'No, Dean,' said Ponder. 'It looks like that, but there are only 21,309 important nodalities at this point. Hex can deal with almost all of them. They involve quite minute changes at the quantum level.'

The wizards continued to stare upwards as the whorls and loops flashed by and dwindled.

'Someone really doesn't want that book,' said the Lecturer in Recent Runes, his face illuminated by the multi coloured glow.

'In theory there isn't a someone in charge,' said Ponder.

'But the odds against Darwin writing *Origin* are getting bigger by the minute!'

'The odds against anything actually happening are huge, when you come to think about it,' said Ridcully. 'Take poker, for example. The odds against four aces are huge, but the odds of having any four cards at all are *really* big.'

'Well put, Archchancellor!' said Ponder. 'But this is a crooked game.'

Ridcully strode out into the centre of the Great Hall, his face illuminated by the glowing map.

'Gentlemen!' he bellowed. 'Some of you already know what this is about, eh? We're going to *force* a history on Roundworld! It's one that should be there already! Something is trying to *kill* it, gentlemen. So if someone wants to stop it happening, we want to make it happen all the more! You will be sent into Roundworld with a series of tasks to do! Most of them have been made very simple so that wizards can understand them! Shortly our missions for tomorrow, should you chose to accept them, will be given to you by Mr Stibbons. If you do *not* choose to accept them, you are free to choose dismissal! We're starting at dawn! Dinner, Second Dinner, Midnight Snack, Somnambulistic Nibbles and Early Breakfast will be served in the Old Refectory! There will be *no* Second Breakfast!'

Over a chorus of protest he went on: 'We are taking this *seriously*, gentlemen!'

TWELVE

THE
WRONG BOOK

OUR FICTIONAL DARWIN HAS A lot more in common with the 'real' one – the Darwin of the particular timeline that you inhabit, the one who wrote *The Origin* and not *The Ology* – than might at first be apparent. Or plausible. The irresistible force of narrativium induces us to imagine Charles Darwin as an old man with a beard, a stick, and a faint but definite hint of gorilla. And so he was, in his old age. But as a young man he was vigorous, athletic, and engaged in the kind of exuberant and not always politically correct activities that we expect of young men.

We've already learned of the real Darwin's amazing fortune in getting on board the *Beagle* and remaining there, culminating in his boundless delight at the geology of the coral island of St Jago. But there are other crucial nodalities, points of intervention, and thaumic occlusions in that version of Roundworld's historical record, and the wizards are exercising extreme care and attention in the hope of steering history through, past, and around these causal singularities.

For example, the *Beagle* really did come under fire from a cannon. When the ship tried to enter the harbour at Buenos Aires in 1832, one of the local guard ships fired at it. Darwin was convinced that he heard a cannonball whistle over his head, but it turned out that the shot was a blank, intended as a warning. FitzRoy, muttering angrily about insults to the British flag, sailed on, but was stopped by a quarantine boat: the harbour authorities were worried about cholera. Incensed, FitzRoy loaded all of the cannons on one side of his ship. As he sailed back out of the harbour he aimed them all at the guard ship, informing its crew that if they ever fired at the *Beagle* again, he would send their 'rotten hulk' to the seabed.

Darwin really did learn to throw a bolas, too, on the pampas of Patagonia. He enjoyed hunting rheas, and watching the gauchos bring them down by entangling a bolas in their legs. But when he tried to do the same, all he managed was to trip up his own horse. *The Origin* might have vanished from history's timeline then and there, but Darwin survived, with only his pride hurt. The gauchos found the whole thing hugely amusing.

Charles even took part in suppressing an insurrection. When the *Beagle* reached Montevideo, shortly after the cannonball incident, FitzRoy complained to the local representative of Her Majesty's Royal Navy, who promptly set sail for Buenos Aires in his frigate HMS *Druid* to secure an apology. No sooner had the warship disappeared from view than there was a rebellion, with black soldiers taking over the town's central fort. The chief of police asked FitzRoy for help, and he dispatched a squad of fifty sailors, armed to the teeth . . . with Darwin happily bringing up the rear. The mutineers immediately surrendered, and Darwin expressed disappointment that not a shot had been fired.

No expense, then, has been spared to bring you historical truth, inasmuch as so weighty a characteristic as truth can be attributed to something as ethereal as history. Except for the giant squid, of course. That happened in a different timeline, when the malign forces were getting extremely desperate and strayed into *Twenty Thousand Leagues Under the Sea* through some obscure warp in L-space.

The most important similarity between the two Darwins is less exciting, but essential to our tale. The real Charles Darwin, like his fictional counterpart, began by writing *the wrong book*. In fact, he wrote eight wrong books. They were very *nice* books, very worthy ... of great scientific value ... and they did his reputation no harm at all ... but they weren't about natural selection, his term for what later scientists would call 'evolution'. Still, that book was brewing merrily away in the back of his mind, and until he was ready to bring it off the back burner, he had plenty of other things to write about.

It had been FitzRoy who had put the idea of authorship into his head. The *Beagle*'s captain had signed himself up to write the story of his round-the-world voyage, based on the ship's log. He had also agreed to edit an accompanying book about a previous survey by the same vessel – the one where Stokes had shot himself. As the *Beagle* headed north-west from Cape town, stopped briefly at Bahía in Brazil, and turned north-east across the Atlantic towards its final destination in Falmouth, FitzRoy suggested to Darwin that the latter's diary might form the basis of a third volume on the natural history of the voyage, completing the trilogy.

Darwin was nervous but excited at the prospect of becoming an author. He had another book in mind, too, on geology. He'd been thinking about it ever since his revelation on the island of St Jago.

As soon as the ship had returned to England, FitzRoy got married and went on honeymoon, but he also made an impressive start to his book. Darwin began to worry that his own slow writing would delay the whole project, but FitzRoy's early enthusiasm soon ground to a halt. Between January and September 1837 Charles worked flat out, overtook the captain, and towards the year's end he sent his finished manuscript to the printer's. It took FitzRoy more than a year to catch up, so Darwin's contribution was held back, finally seeing the light of day in 1839 as volume 3 of the *Narrative of the Surveying Voyages of H.M.S. Adventure and Beagle, Between the Years 1826 and 1836*, with the subtitle *Volume 3: Journal and Remarks, 1832–1836*. After a few months the publishers reissued it on its own as *Journal of Researches into the Geology and Natural History of the Various Countries Visited by H.M.S. Beagle*. It may have been the wrong book, but writing it had one very useful effect on Darwin's thinking. It forced him to try to make sense of all the things he had seen. Was there some overarching principle that could explain it all?

Next came his geology book, which eventually turned into three: one on coral reefs, one on volcanic islands, and one on the geology of South America. These established his scientific credentials and led to him winning a major Royal Society prize. Darwin was now recognised as one of the leading scientists in the land.

He was also making ever more extensive notes on the transmutation of species, but he still was in no hurry to publish. Quite the contrary. Elsewhere, political forces were at work aiming to destroy the influence of the Church, and one of their key points was that living creatures could easily have arisen without the intervention of a creator. Darwin, being (at that point in his life) a good Christian, was totally averse to anything that might seem to ally him with such people. He could not publicly espouse transmutation without risking serious damage to the Anglican Church, and nothing in the world would induce him to contemplate that. But his deep insight about natural selection wouldn't go away, so he continued developing it as a kind of hobby.

He did mention the insight to various scientific friends and acquaintances, among them Lyell, and also Joseph Dalton Hooker, who didn't dismiss the idea out of hand. But he did tell Darwin, 'I shall be delighted to hear how you think this change may have taken place, as no presently conceived opinions satisfy me on this subject.' And he later said, rather acerbically, that 'No one has hardly a right to examine the question of species who has not minutely examined many.' Darwin took this advice to heart and cast around for new species to become an expert on. In 1846 he sent the final proofs of his geology books back to the printer and celebrated by collecting the last bottle of preserved specimens from the *Beagle* voyage. At the top of the bottle he noticed a crustacean from the Chonos Archipelago – a barnacle.

That would do. It was as good as anything else.

Hooker helped Darwin set up his microscope and make some preliminary anatomical observations. Darwin asked Hooker to name the new beast, and together they decided on *Arthrobalanus.** 'Mr Arthrobalanus', as they privately called it, turned out to be somewhat unusual. 'I believe *Arthrobalanus* has no ovisac at all!' Charles wrote. 'The appearance of one is entirely owing to the splitting & tucking up of the posterior penis.' To resolve the mystery he took other barnacles from the bottle and looked at them, too. Now he was doing comparative anatomy of barnacles, and enjoying the hands-on experience immensely. This was better than writing.

By Christmas he had decided to study every barnacle known to humanity – the entire order of *Cirripedia*. Which turned out to be rather a lot, so he settled for the British ones. Even these were rather a lot, and in the end the task took eight years.

He might have finished earlier, but in 1848 he got interested in barnacle sex, and that was very peculiar indeed. Most barnacles were hermaphrodites, able to assume either sex. But some species had good old-fashioned males and females. Except that the males spent much of their lives *embedded in the females*.

Not only that: some supposedly hermaphrodite species also had tiny males that somehow assisted in the reproductive process.

Now Darwin became very excited, because he had convinced himself that what he was observing was a relic of evolution, as a hermaphrodite ancestor gradually developed separate sexes. A 'missing link' for barnacle sex. He could reconstruct the barnacles' family tree, and what he thought he saw reinforced his views on natural selection. So even when he tried to do respectable science, and become a taxonomist, transmutation insisted in getting in on the act. In fact, if anything convinced Darwin he was right about transmutation, it was barnacles.

---

*. Literally, 'jointed acorn'

He became ill, but continued working on barnacles. In 1851 he published two books about them – one on fossil barnacles for the Palaeontographical Society, the other on the living ones for the Royal Society. By 1854 he had produced a sequel to each of them.

These were Darwin's eight wrong books:

1839   *Journal of Researches into the Geology and Natural History of the Various Countries Visited by H.M.S. Beagle*

1842   *The Structure and Distribution of Coral Reefs*

1844   *Geological Observations on the Volcanic Islands Visited During the Voyage of H.M.S. Beagle*

1846   *Geological Observations of South America*

1851   *A Monograph on the Fossil Lepadidae, or, Pedunculated Cirripedes of Great Britain*

1851   *A Monograph on the Sub-class Cirripedia volume 1*

1854   *A Monograph on the Fossil Balanidae and Verrucidae of Great Britain*

1854   *A Monograph on the Sub-class Cirripedia volume 2*

Not a hint of transmutation of species, the struggle for life, or natural selection.

Yet, in a strange way, all of his books – even the geological ones – were crucial steps towards the work that was now putting itself together inside his head. Darwin's ninth book would be pure dynamite. He wanted desperately to write it, but he had already decided that it would be far too dangerous to be published.

It is a common dilemma in science: whether to publish and be damned, or not to publish and be pre-empted. You can have the credit for a truly revolutionary idea, or a quiet life, but not both.

Darwin was wary of publicity, and he was scared that putting his views into print might damage the Church. But there is nothing that more effectively galvanises a scientist than the fear that somebody else will pip them to the winning post. In this case, that somebody was Alfred Russel Wallace.

Wallace was another Victorian explorer, equally keen on natural history. Mostly because he could sell it. Unlike Darwin, he was not 'gentry', and had no independent income. He was the son of an impecunious lawyer* and had been taken on at age fourteen as a builder's apprentice. He spent his evenings drinking free coffee in the Hall of Science off Tottenham Court Road in London. This was a socialist organisation, dedicated to the overthrow of private property and the downfall of the Church. Wallace's experiences as a youth reinforced a left-wing view of politics. He financed his own travels, and made a living by selling the specimens he collected – butterflies, beetles (a thousand labelled specimens per box, the dealers demanded†), even bird skins. He went on a collecting expedition to the Amazon in 1848, and again to the Malay Archipelago in 1854. There, in Borneo, he sought orang-utans. The idea that humans were somehow related to the great apes was simmering away in the collective subconscious, and Wallace wanted to investigate a potential human ancestor.‡

One miserable Borneo day, when a tropical monsoon raged outside and Wallace was stuck indoors, he put together a little scientific paper outlining some modest ideas that had just popped into his head. It eventually appeared in the *Annals and Magazine of Natural History*, a rather ordinary publication, and it was about the 'introduction' of species. Lyell, aware of Darwin's secret interest in such matters, pointed the paper out to him, and Charles began to read it. Then another of Charles's regular correspondents, Edward Blyth,

---

* Yes, we *know* it sounds unlikely, but apparently there are such things.
† It was a good job that God had such a fondness for beetles.
‡ Its potential for Librarianship was not widely recognised at that time.

wrote from Calcutta with the same recommendation. 'What do you think of Wallace's paper in the Ann M.N.H.? Good! Upon the whole!' Darwin had met Wallace shortly before one of the latter's expeditions – he couldn't remember which – and he could see that the Ann M.N.H. paper had useful things to say about relationships between similar species. Especially the role of geography. But apart from that, he felt that the paper contained nothing new, and made an entry to that effect in one of his notebooks. Anyway, it seemed to Darwin that Wallace was talking about creation, not evolution. Nevertheless, he wrote to Wallace, encouraging him to continue developing his theory.

This was a Really Bad Idea.

Encouraged by Lyell and others, who were now warning him that if he delayed too long, others might snatch the prize, Darwin was putting together ever more elaborate essays on natural selection, but he continued to dither about publication. All that changed in an instant in June 1858, when the postman dropped a bombshell through Charles's letterbox. It was a package from Wallace, containing a twenty-page letter, sent from the Moluccas. Wallace had taken Darwin's advice to heart. And he had come up with a very similar theory. Very similar indeed.

Calamity. Darwin declared that his life's work was 'smashed'. 'Your words have come true with a vengeance,' he wrote to Lyell. The more he read Wallace's notes, the closer the ideas seemed to his own. 'If Wallace had my MS [manuscript] sketch written out in 1842, he could not have made a better short abstract!' Darwin moaned in a letter to Lyell.

Staid Victorians would soon consider both Wallace and Darwin to be out of their minds, and Wallace certainly came close, for he was suffering from malaria when he composed his letter to Darwin. As a good socialist, Wallace had been taught not to trust the reasoning of Malthus, who had argued that the world's ability to feed itself grew linearly, while the population grew exponentially – implying that eventually the population would win and there would be too little food to go round. Socialists believed that human ingenuity could postpone such an event indefinitely. But by the 1850s even socialists were beginning to view Malthus in a more favourable light; after all, the threat of overpopulation was a very good reason to promote contraception, which made excellent sense to every good socialist. Half-delirious with fever, Wallace thought about the rich variety of species he had encountered, wondered how that fitted in with Malthus, put two and two together, and realised that you could have selective breeding without the need for a breeder.

As it turned out, he didn't have *quite* the same view as Darwin. Wallace thought that the main selective pressure came from the struggle to survive in a hostile environment – drought, storm, flood, whatever. It was this struggle that removed unfit creatures from the breeding pool. Darwin had a rather blunter view of the selection mechanism: competition among the organisms themselves. It wasn't quite 'Nature red in tooth and claw' as Tennyson had written in his *In Memoriam* of 1850, but the claws were unsheathed and there was a certain pinkness to the teeth. To Darwin, the environment set a background of limited resources, but it was the creatures themselves that selected each other for the chop when they *competed* for those resources. Wallace's political leanings made him detect a purpose in natural selection: to 'realise the ideal of a perfect man'. Darwin refused even to contemplate this kind of utopian hogwash.

Wallace hadn't mentioned publishing his theory, but Darwin now felt obliged to recommend it to him. At that point it looked as if Charles had compounded his Really Bad Idea, but for once the universe was kind. Lyell, searching for a compromise, suggested that the two men might agree to publish their discoveries simultaneously. Darwin was concerned that this might make it look as if he'd pinched Wallace's theory, worried himself to distraction, and finally handed the negotiating over to Lyell and Hooker and washed his hands of it.

Fortunately, Wallace was a true gentleman (the accident of breeding notwithstanding) and he agreed that it would be unfair to Darwin to do anything else. He hadn't realised that Darwin had been working on exactly the same theory for many years, and he had no wish to steal such an eminent scientist's thunder, perish the very thought. Darwin quickly put together a short version of his own work, and Hooker and Lyell got the two papers inserted into the schedule of the Linnaean Society, a relatively new association for natural history. The Society was about to shut up shop for the summer, but the council fitted in an extra meeting at the last minute, and the two papers were duly read to an audience of about thirty fellows.

What did the fellows make of them? The President reported later that 1858 had been a rather dull year, not 'marked by any of those striking discoveries which at once revolutionise, so to speak, our department of science'.

No matter. Darwin's fear of controversy was now irrelevant, because the cat was out of the bag, and there was no chance whatsoever that the beast could be stuffed back in. Yet, as it happened, the anticipated controversy didn't quite materialise. The meeting of the Linnaean Society had been rushed, and the fellows had departed muttering vaguely under their breaths, feeling that they *ought* to be outraged by such blasphemous ideas … yet puzzled because the enormously respected (and respectable) Hooker and Lyell clearly felt that both papers had some merit.

And the ideas struck home with some. In particular, the Vice-President promptly removed all mention of the fixity of species from a paper he was working on.

Now that Darwin had been forced to put his head above the parapet, he would lose nothing by publishing the book that he had previously decided not to write, but had constantly been thinking about anyway. He had intended it to be a vast, multi-volume treatise with extensive references to scientific literature, examining every aspect of his theory. It was going to be called *Natural Selection* (a conscious or subconscious reference to Paley's *Natural Theology?*). But time was pressing. He polished up his existing essay, changing the title to *On the Origin of Species and Varieties by Means of Natural Selection.* Then, on the insistent advice of his publisher John Murray, he cut out the 'and Varieties'. The first print run of 1250 copies went on sale in November 1859. Darwin sent Wallace a complimentary copy, with a note: 'God knows what the public will think.'

In the event, the book sold out before publication. Over 1500 advance orders came in for those 1250 copies, and Darwin promptly started working on revisions for a second edition. Charles Kingsley, author of *The Water-Babies*, country rector, and Christian socialist, loved it, and wrote a lavish letter: '[It is] just as noble a conception of Deity, to believe that He created primal forms capable of self-development . . . as to believe that He required a fresh act of intervention to supply the *lacunas** which He himself had made.' Kingsley was something of a maverick, because of his socialist views, so praise from this source was something of a poisoned chalice.

The reviews, steadfast in their Christian orthodoxy, were distinctly less favourable. Even though *Origin* hardly mentions humanity, all the usual complaints about men and monkeys, and insults to God and His Church, were trotted out. What particularly galled the reviewers was that *ordinary people* were buying the thing. It was all right for the upper classes to toy with radical views, it had an attractive frisson of naughtiness and was perfectly harmless among gentlemen of breeding, though not ladies of course; but those same views might put ideas into the common folk's heads, if they were exposed to them, and upset the established order. For Heaven's sake, the book was even selling to commuters outside Waterloo railway station! It must be suppressed!

---

* No, not long-haired South American beasts of burden, but Latin for 'gaps'.

Too late. Murray geared up to print 3000 copies of the second edition, whose likely sales were not going to suffer from public controversy. And the people who mattered most to Darwin – Lyell, Hooker, and the anti-religious 'evangelist' Thomas Henry Huxley – were impressed, and pretty much convinced. While Charles stayed out of the public debate, Huxley set to with a will. He was determined to advance the cause of atheism, and *Origin* gave him a point of leverage. The radical atheists loved the book, of course: its overall message and scientific weightiness were enough for them, and they weren't too concerned about fine points. Hewett Watson declared Darwin to be 'the greatest revolutionist in natural history of this century'.

In the introduction to *Origin*, Darwin begins by telling his readers the background to his discovery:

> When on board H.M.S. *Beagle*, as naturalist, I was much struck with certain facts, in the distribution of the inhabitants of South America, and in the geological relations of the present to the past inhabitants of that continent. These facts seemed to me to throw some light on the origin of species – that mystery of mysteries, as it has been called by one of our greatest philosophers. On my return home, it occurred to me, in 1837, that something might perhaps be made out of this question by patiently accumulating and reflecting on all sorts of facts which could possibly have any bearing on it.

Apologising profusely for lack of space, and time, to write something more comprehensive than his 150,000-word tome, Darwin then moves towards a short summary of his main idea. Writers on science generally appreciate that it is seldom enough to discuss the *answer* to a question: it is also necessary to explain the question. And that, of course, should be done first. Otherwise your readers will not appreciate the context into which the answer fits. Darwin was clearly aware of this principle, so he begins by pointing out that:

> It is quite conceivable that a naturalist, reflecting on the mutual affinities of organic beings, on their embryological relations, their geographical distribution, geological succession, and other such facts, might come to the conclusion that each species had not been independently created, but had descended, like varieties, from other species. Nevertheless such a conclusion, even if well founded, would be unsatisfactory, until it could be shown how innumerable species inhabiting this world have been modified, so as to acquire those perfections of structure and coadaptation which most justly excites our admiration.

Already we see a gesture towards Paley – 'perfections of structure' is a clear reference to the watch/watchmaker argument, and 'had not been independently created' shows that Darwin doesn't buy Paley's conclusion. But we also see something that characterises the whole of *Origin*: Darwin's willingness to acknowledge difficulties in his theory. Time and again he raises possible objections – not as straw men, to be knocked flat again, but as serious points to be considered. More than once he concludes that there is more to be learned, before the objection can be resolved. Paley, to his credit, did something similar, though he didn't go as far as admitting ignorance: he knew that he was right. Darwin, a real scientist, not only had his doubts – he shared them with his readers. He would not have arrived at his theory to begin with if he had failed to seek the weaknesses of the hypotheses upon which it was based.

He also, of course, makes it clear what his own work is adding to the speculations of earlier 'transmutationists'. Namely: he has come up with a *mechanism* for species change. There are advantages in being honest about your own limitations: you gain the right to talk about the limitations of others. And now he tells us what that mechanism is. Species, we know, are variable – the domestication of wild species like chickens, cows, and dogs is clear evidence of that. Although that is deliberate selection by humans, it opens the door to selection by nature without human aid:

> I will then pass on to the variability of species in a state of nature . . . We shall, however, be enabled to discuss what circumstances are most favourable to variation. In the next chapter the Struggle for Existence amongst all organic beings throughout the world, which inevitably follows from their high geometrical powers of increase, will be treated of . . . The fundamental subject of Natural Selection will be treated at some length in the fourth chapter; and we shall then see how Natural Selection almost inevitably causes much Extinction of the less improved forms of life, and induces what I have called Divergence of Character.

He then promises four chapters on 'the most apparent and gravest difficulties of the theory', prominent among these being to understand how a simple organism or organ can change into a highly complex one – another nod to Paley. The introduction ends with a flourish:

> I can entertain no doubt . . . that the view which most naturalists entertain, and which I formerly entertained – namely, that each species has been independently created – is erroneous. I am fully convinced that species are not immutable; but that those belonging to what are called the same genera are lineal descendent of some other and generally extinct species . . . Furthermore, I am convinced that Natural Selection has been the main but not exclusive means of modification.

In essence, Darwin's theory of natural selection, which soon became known as evolution,* is straightforward. Most people think they

---

* The term was around in Victorian times, as a phenomenon but not a specific mechanism. Darwin didn't use it in *Origin*, nor in the later *The Descent of Man*. However, the final word in *Origin* is 'evolved'.

understand it, but its simplicity is deceptive, and its subtleties are easily underestimated. Many of the standard criticisms of evolutionary theory stem from common misunderstandings, not from what the theory actually proposes. The ongoing scientific debate about details is often misrepresented as disagreement with the general outline, which is an error based on too simple-minded a view of how science develops and what 'knowledge' is.

Briefly, Darwin's theory goes like this.

1. Organisms, even those in the same species, are variable. Some are bigger than others, or bolder than others, or prettier than others.
2. This variability is to some extent hereditary, passed on to offspring.
3. Unchecked population growth would quickly exhaust the capacity of the planet, so something checks it: competition for limited resources.
4. Therefore as time passes, the organisms that *do* survive long enough to breed will be modified in ways that *improve* their chance of surviving to breed, a process called natural selection.
5. Ongoing slow changes can lead, in the long run, to big differences.
6. The long run has been very long indeed – hundreds of millions of years, maybe more. So by now those differences can have become huge.

It's relatively simple to put these six ingredients together and deduce that new species can arise without divine intervention – provided we can justify each ingredient.

Even though different species seem to stay pretty much the same – think lions, tigers, elephants, hippos, whatever – it is actually rather obvious that, in general, species are not fixed for all time. The changes are relatively slow, which is why we don't notice them. But they do happen. We've already seen that in Darwin's finches, evolutionary changes can be and have been observed on a timescale of years, and in bacteria they occur on a timescale of days.

The most obvious evidence for the variability of species, in Darwin's day and ours, was the domestication of animals – sheep, cows, pigs, chickens, dogs, cats . . .

. . . and pigeons. Darwin was rather knowledgeable about pigeons, he belonged to two London pigeon clubs. Every pigeon-fancier knows that by selectively breeding particular combinations of male and female pigeons, it is possible to produce 'varieties' of pigeons with particular characteristics. 'The diversity of the breeds is something astonishing,' says Darwin in the first chapter of *Origin*. The English carrier pigeon has a wide mouth, large nostrils, elongated eyelids, a long beak. The short-faced tumbler has a short stubby beak like a finch. The common tumbler flies high up in a tightly knit flock, and has an odd habit of falling about in the sky, whence its name. The runt (despite its name) is huge, with a long beak and large feet. The barb is like the carrier but with a short, broad beak. The pouter has an inflatable crop and can puff out its chest. The turbit has a short beak and a line of reversed feathers on its chest. The Jacobin has so many reversed feathers on its neck that they form a hood. Then there are the trumpeter, laugher, fantail . . . These are not separate species: they can interbreed, to produce viable 'hybrids' – cross-breeds.

The enormous variety of dogs is so well known that we don't even need to mention examples. It's not that the dog species is exceptionally malleable, just that dog-breeders have been unusually active and imaginative. There is a dog for every purpose that a dog can carry out. Again, they're all *dogs*, not new (albeit related) species. They can *mostly* (barring really big size differences) interbreed, and artificial insemination can take care of mere size. Dog sperm plus dog egg makes fertile dog zygote, and, eventually, dog – independently of breed. That's why pedigree pooches need a pedigree, to guarantee that their parentage is pure. If the different varieties of dog were different species, that wouldn't be necessary.

In modern times, it has become clear that cats are just as malleable, but the cat-breeders have only just got going on exotic cats. The same goes for cows, pigs, goats, sheep . . . and what about flowers? The number of varieties of garden flowers is immense.

By avoiding the creation of hybrids, the breeder can maintain the individual varieties over many generations. Pouter pigeons breed with pouters to produce (a substantial proportion of) pouters. Carriers mated with carriers produce (mostly) carriers. The underlying genetics, about which Darwin and his contemporaries knew nothing, is complicated enough that apparent hybrids can sometimes arise from what seems to be pure stock, just as two brown-eyed parents can nonetheless have a blue-eyed child. So pigeon-breeders have to eliminate the hybrids.

The existence of these cross-bred varieties does not, of itself, explain how new species can arise of their own accord. Varieties are not species; moreover, the guiding hand of the breeder is evident. But varieties do make it clear that there must be plenty of variability within a species. In fact, the variability is so great that one can readily imagine selective breeding leading to entirely new species, given enough time. And the avoidance of hybrids can maintain varieties from one generation to the next, so their *characters* (biologese for the features that distinguish them) are *heritable* (biologese for 'able to be passed from one generation to the next'). So Darwin has his first ingredient: heritable variability.

The next ingredient was easier (though still controversial in some quarters). It was time. Oodles and oodles of time, the Deep Time of geologists. Not a few thousand years, but millions, tens of millions . . . billions, in fact, though that was further than the Victorians were willing to go. Deep Time, as we've previously observed, is contrary to the biblical chronology of Bishop Ussher, which is why the idea remains controversial among certain Christian fundamentalists, who have bizarrely chosen to fight their corner on the weakest of grounds, completely needlessly. Deep Time is supported by so much evidence that a truly committed fundamentalist has to believe that his God is deliberately trying to fool him. Worse, if we can't trust the evidence of our own eyes, then we can't trust the apparent element of 'design' in living creatures either. We can't trust anything.

Lyell realised that the age of the Earth must be many millions of years, when he looked at sedimentary rocks. These are rocks like limestone or sandstone which form in layers, and have been deposited either underwater, as muddy sediments, or in deserts, as accumulating sand. (Independent evidence for these processes comes from the fossils found in such rocks.) By studying the rate at which modern sediments accumulate, and comparing that with the thickness of known beds of sedimentary rock, Lyell could estimate the time it had taken for the layers of rock to be deposited. Something in the range 1000–10,000 years would produce a layer about a metre thick. But the chalk cliffs of the south coast, around Dover, are hundreds of metres thick. So that's several hundred thousand years of deposition, and we've only dealt with one of the numerous layers of rock that make up the geological column – the historical sequence of different rocks.

We now have many other kinds of evidence for the great age of our planet. The rate of decay of radioactive elements, which we can measure today and extrapolate backwards, is in general agreement with the evidence of the rock layers. The rate of movement of the continents, when combined with the distances they have moved, is again consistent with other estimates. We've seen that India was once attached to Africa, but about 200 million years ago it broke off, and by 40 million years ago it had moved all the way to its current position, butting up against Asia and pushing up the Himalayas.

When continents move apart – as Africa and South America, or Europe and North America, are doing now – new material forms on the ocean floor, flowing out from the mantle beneath to form huge mid-ocean ridges. The rocks in the ridges contain a record of the changes in the Earth's magnetic field, 'frozen in' as the rock cooled. They show a long series of repeated reversals of the field polarity. Sometimes the 'north' magnetic pole is at the northern end of the Earth, as now, but every so often the polarity flips, so that the magnetic pole near the northern end is the 'south' one. Mathematical models of the Earth's magnetic field predict that such reversals occur roughly once every five million years. Count the number of reversals in the ocean-ridge rocks, multiply by five million . . . again, the numbers fit reasonably well, and careful comparisons and a lot of disputation by experts lead to revised numbers that fit even better.

The Grand Canyon is a deep gash through layers of rock one mile (1.6km) thick. You have a choice. You can understand what the record of the rocks is telling you here: it took a very long time to lay down those rocks, and quite a long time – though less – for flash-flooding in the Colorado river to erode them again. Or you can follow one book that until recently was displayed in the 'science' section of the Grand Canyon bookstore, until a lot of scientists complained, and assert that the Grand Canyon is evidence for Noah's flood. The first choice fits huge amounts of evidence and geological understanding. The second is an excellent test of faith, because it fits absolutely nothing. A flood that lasted only 40 days could never have produced that kind of geological formation. A miracle? In that case, the Sahara desert could equally well be hailed as evidence for Noah's flood, miraculously *not* forming a deep canyon. Once you admit miracles, you can't pursue a logical thread.

Anyway, that's the second ingredient – Deep Time. It takes huge amounts of time to change organisms into entirely new species, if all you can do – as Darwin believed – is make very gradual changes. But even Deep Time, when combined with heritable variation, is not enough to lead to the kind of organised, coherent changes that are needed to create new species. There has to be a *reason* for such changes to occur, as well as opportunity and time. Darwin, as we've seen, found his reason in Malthus's contention that the unchecked growth of organisms is exponential, whereas that of resources is linear. In the long run, exponential growth always wins.

The first assertion is pretty much correct, the second highly debatable. The qualifier 'unchecked' is crucial, and real populations only grow exponentially if there are plenty of resources available. Typically, the growth starts exponentially with a small population and then levels off as the population size increases. But in most species, two parents (let's think sexual species here) produce some larger number of offspring. A breeding female starling lays about 16 eggs in her life, and with 'unchecked' growth, the starling population would multiply by 8 every lifetime. It would not be long before the planet was knee-deep in starlings. So, of necessity, 14 of those 16 offspring (on average) fail to breed – usually because something eats them. Just two become parents in their turn. A female frog may lay 10,000 eggs in her life, and nearly all die in various grotesque ways to achieve each two parents; a female cod contributes forty million or thereabouts of her offspring to planktonic food chains, for each two that breed. Here the multiplier, with 'unchecked' growth, would be 20 million per cod-lifetime. Unchecked growth simply doesn't bear thinking about as a realistic prospect.

We suspect that Malthus plumped for linear growth of resources for a slightly silly reason. Victorian school-textbook mathematics distinguished two main types of sequence: geometric (exponential) and arithmetic (linear). There were plenty of other possibilities, but they didn't get into the textbooks. Having already assigned geometric growth to organisms, Malthus was left with arithmetic growth for resources. His main point doesn't depend on the actual growth rate, in any case, as long as it is less than exponential. As the starling example shows, most offspring die before breeding, and that's the main point here.

Given that most young starlings cannot possibly become parents, the question arises: which ones will? Darwin felt that the ones that survived to breed would be the ones best suited for survival, which makes sense. If one starling is better at finding food, or hanging on to it, than another one, then it's clear which one is more likely to do best if food supplies become limited. The better one *might* be unlucky and get eaten by a hawk; but across the population, starlings that are better equipped to survive are generally the ones that *do* survive.

This process of 'natural selection' in effect plays the role of an external breeder. It *chooses* certain organisms and eliminates the rest. The choice is not conscious – there is no consciousness to do the choosing, and no preconceived *purpose* – but the end result is very similar. The main difference is that natural selection makes sensible choices, whereas human selection can make ridiculous ones (like dogs with faces so flattened they can hardly breathe). Sensible choices lead to sensible animals and plants, ones that are beautifully adapted for survival in whatever environment they happened to be in when natural selection was moulding them.

It is just like breeding new varieties of pigeon, but without a human breeder. Natural selection exploits the same variability of organisms that pigeon-breeding does. It makes choices based on survival value (in some environment) rather than whim. It is typically much slower than human intervention, but the timescale is so vast that this slowness doesn't matter much. Heritable variation plus natural selection inevitably lead, over Deep Time, to the origination of species.

Nature does it all on her own. There is no need for a series of acts of special creation. That doesn't imply that special creation has not occurred. It just removes any logical imperative for it.

Paley was wrong.

The watches don't need a watchmaker.

They can make themselves.

THIRTEEN

INFINITY IS
A BIT TRICKY

 IT WAS JUST GONE HALF five in the morning, too late for Nibbles and yet not time for Early Breakfast. Jogging through the grey mist, Archchancellor Ridcully saw the lights on in the Great Hall. Steeling himself in case Ponder had students in there, he pushed open the door.

There were a few students around. One of them was asleep under the coffee spigot.

Ponder Stibbons was still on the stepladder, waving his hands through the timelines.

'Getting anywhere, Stibbons?' said Ridcully, running on the spot.

Ponder managed to steady himself just in time.

'Er . . . general progress, sir,' he said, and climbed down.

'Bit of a big job, eh?' said Ridcully.

'Rather taxing, sir, yes. We've done the instructions, though. We're nearly ready.'

'Hit 'em hard, that's the style,' said Ridcully, punching the air.

'Quite probably, sir,' said Ponder, yawning.

'I was thinking while I was running, Stibbons, as is my wont,' said Ridcully.

It's going to be about the eyeball, isn't it, Ponder thought. I'm pretty good on the eyeball now, but then he'll ask about the parasitic wasp and that's a puzzler, and then he'll ask how *exactly* is evolution passed on and there's a god-space right there. And then he'll ask *how* do you get from a blob in the ocean to people by adding nothing but sunlight and time? And he'll probably say: people know they're people, did blobs know they were blobs? What bit of a blob knows that? Where did consciousness come from, then? Did the big lizards have it? What's it for? What about imagination? And even if I can think up some kind of answers to all those, he'll say: look, Stibbons, what you've got there is a lot of clockwork answers, and if I ask you how you can get from a big bang to turtles and spoons and Darwin, all you'll be able to come up with is more clockwork. *How* did all this happen? Who wound it up? How can nothing explode? *Theology of Species* makes so much sense when—

'Are you all right, Stibbons?' He was aware of the Archchancellor looking at him with uncharacteristic concern.

'Yes, sir, Just a bit tired.'

'Only, your lips were moving.'

Ponder sighed. 'What was it you were thinking about, sir?'

'Lots of Darwins get through this voyage, right?'

'Yes. An infinite number.'

'Well, in that case—' the Archchancellor began.

'But Hex did say it's a much smaller infinite number that the number that don't,' said Ponder. 'And that's an even smaller number than the very large infinity when he never goes on the voyage. And the number of infinities where he's never even born is—'

'Infinite?' Ridcully asked.

'At least,' said Ponder. 'However, there is a positive side to this.'

'Do tell, Stibbons.'

'Well, sir, once *Origin* is published, the number of universes in which it is published will also become infinite in an infinitely small space of time. So even though the book may only be written once, it will, by human standards, immediately have been written in untold billions of adjacent universes.'

'An infinite number, I suspect?' said Ridcully.

'Yes, sir. Sorry about that. Infinity is a bit tricky.'

'You can't imagine half of it, for one thing.'

'That's true. It's not really a number at all. You can't get to it start-ing from one. And that's the problem, sir. Hex is right, the oddest number in the multiverse isn't infinity, it's one. Just one Charles Darwin writing *The Origin of Species* . . . it's impossible.'

Ridcully sat down. 'I'll be damn glad when he finishes the book,' he said. 'We'll get all those nody things sorted and get him back and I personally will hand him the pen.'

'Er . . . that doesn't happen immediately, sir,' said Ponder. 'He didn't write it until he was back home.'

'Fair enough,' said Ridcully. 'Probably a bit tricky, writin' on a boat.'

'He thought about it a lot first, sir,' said Ponder. 'I did mention that.'

'How long?' said Ridcully.

'About twenty-five years, sir.'

'What?'

'He wanted to be sure, sir. He researched and wrote letters, lots of letters. He wanted to know everything about, well, everything – silk worms, sheep, jaguars . . . He wanted to be sure he was right.' Ponder thumbed through the papers on his clipboard. 'This inter-ested me. It was from a letter he wrote in 1857, and he says "*what a jump it is from a well-marked variety, produced by a natural cause, to a species produced by the separate act of the Hand of God*".

'That's the author of *The Origin?* Sounds more like the author of *The Ology.*'

'It was a big thing he was going to do, sir. It worried him.'

'I've read *The Ology*,' said Ridcully. 'Well, some of it. Makes a lot of sense.'

'Yes, sir.'

'I mean, if we hadn't watched the world all happen from Day One, we'd have thought—'

'I know what you mean, sir. I think that's why *The Ology* was so popular.'

'Darwin – I mean *our* Darwin – thought that no god would make so many kinds of barnacle. It's so wasteful. A perfect being wouldn't do it, he thought. But the other Darwins, the religious ones, said that was the whole point. They said that just as mankind had to strive for perfection, so must the whole animal kingdom. Plants, too. Survival of the Worthiest, they called it. Things weren't made perfect, but had an inbuilt, er, striving to achieve perfection, as if part of the Plan was inside them. They could evolve. In fact, that was a good thing. It meant they were getting better.'

'Seems logical,' said Ridcully. 'By god logic, at least.'

'And there's the whole thing about the Garden of Eden and the end of the world,' said Ponder.

'I must've missed that chapter,' said Ridcully.

'Well, sir, it's your basic myth of a golden age at the start of the world and terrible destruction at the end of it, but codified in some very interesting language. Darwin suggested that the early chroniclers had got things mixed up. Like trolls, you know? They think the past is ahead of them because they can see it? The terrible destruction was in fact the *birth* of the world—'

'Oh, you mean the red hot rocks, planets smacking together, that sort of thing?'

'Exactly. And the *end* of the world, well, as experienced, would be the assembly of perfect creatures and plants in a perfect garden, belonging to the god.'

'To get congratulated, and so on? Prizes handed out, marks awarded?'

'Could be, sir.'

'Like an everlasting picnic?'

'He didn't put it like that, but I suppose so.'

'What about the perfect wasps?' said Ridcully. 'You always get them, you know. And ants.'

Ponder had been ready for this.

'There was a lot of debate about that sort of thing,' he said.

'And it concluded how?' said the Archchancellor.

'It was decided that it was the kind of subject on which there could be a lot of debate, and that earthly considerations would not apply.'

'Hah! And Darwin got all this past the priests?'

'Oh, yes. Most of them, anyway,' said Ponder.

'But he was turning their whole world upside down!'

'Um, that was happening anyway, sir. But *this* way, the god didn't drop out of the bottom. People were poking around and proving that the world really was very old, that seabeds had become mountain tops, that all kinds of strange animals had lived a long, long time ago. Lots of people already accepted the idea of evolution. The idea of natural selection, as Darwin called it, of life just evolving itself, was hovering in the air. It was a big threat. But *Theology of Species* said there was a Plan. A huge, divine Plan, unfolding across millions of years! It even included the planet itself! All that turmoil and volcanoes and drowning lands, that was a world evolving, you see? A world that would end up with topsoil, and the right kind of atmosphere, and minerals that were easily accessible, and seas full of fish—'

'A world for humans, in other words.'

'Got it in one, sir,' said Ponder. 'Humans. The top of the tree. A creature that knew what it was, that gave things names, that had a concept of epiphany. That Darwin later wrote another book, called *The Ascent of Man*. Oddly enough, *our* Darwin is going to write a similar book called *The Descent of Man*—'

'Ah, I can see a bad choice of words right there,' said Ridcully.

'Quite,' said Ponder. 'The *Ology* Darwin was considered daring but . . . acceptable. And there was so much evidence that this *was* a planet made for humans. The religion changed quite a lot, but so did the technomancy. The god was still in charge.'

'All very neat,' said the Archchancellor. 'So . . . what about the dinosaurs?'

'Sorry, sir?'

'Mr Stibbons, you *know* what I'm talking about. We saw them, remember? Not the big ones, the little ones who painted their bodies and herded animals? And the octopuses building cities under the sea? Not to mention the crabs! Oh, yes, the crabs. They were really doing well, the crabs. They were building rafts with sails and enslaving other crab nations. That's practically civilisation! But they all got wiped out. Was *that* part of a divine plan?'

Ponder hesitated.

'They did worship a crab-shaped god,' he said, as a holding statement until actual thought happened.

'Well, they would, wouldn't they?' said Ridcully. 'They were *crabs*.'

'Um. Perhaps they just weren't ... satisfactory?' said Ponder. 'In some way?'

'They were pretty clever,' said Ridcully.

Ponder squirmed. 'Darwin didn't know about them,' he said. 'They didn't build anything that lasted. I suppose the Darwin who wrote *The Ology* would have taken the view that they simply failed, or were wicked in some way. One of the major religious texts does mention a divine flood that drowned everything in the world except one family and a boatload of animals.'

'Why?'

'Because they were all wicked, I believe.'

'How can animals be wicked? How can a crab be wicked, for that matter?'

'I don't know, Archchancellor!' Ponder burst out. 'Maybe if they eat forbidden seaweed? Dig a burrow on the wrong day? I'm not a theologian!'

They sat in despondent silence.

'It's a bit of mess, isn't it?,' said Ridcully.

'Yes, sir.'

'We've really got to see to it that *The Origin* gets written.'

'We have indeed, sir.'

'But you'd like to think there's someone in charge, yes?' said

Ridcully, gently. 'Of everything, I mean.'

'Yes! Yes, I would, sir! Not a big beard in the sky, but . . . something! Some kind of frame, some sense that good and bad have real meanings! I can see why *The Ology* was so popular. It wrapped everything up! But how does evolution get passed on? Where does order come from? If you start with a lot of exploding firmament, *how* do you end up with butterflies? Were butterflies built in from the start? How? What bit of burning hydrogen carried the plans for people? Even the Darwin who wrote *The Origin* called on a god to *start* life. It's be nice to know that underneath it all is some kind of . . . sense.'

'You didn't used to talk like this, Mr Stibbons.'

Ponder sagged. 'Sorry, sir. It's all getting me down, I think.'

'Well, I can see why,' said Ridcully. 'Surely there *must* be some Deitium here. Some things can't just happen. Now, the eyeball –'

Ponder gave a little yelp.

'– is easy,' said Ridcully. 'Are you all right, Stibbons?'

'Er, fine, fine, sir. I'm fine. Easy, is it?'

'Seeing keeps you alive,' said Ridcully. 'Any kind of seeing is better than nothing. I can see, ha, what the *Origin* Darwin is getting at there. You don't have to have a god. But there's a kind of wasp that's parasitical on a spider . . . unless I'm thinking about a kind of spider that is parasitical on a wasp . . . anyway, what it does is, it waits until—'

'Ah,' said Ponder brightly, 'wasn't that the gong for Early Breakfast?'

'*I* didn't hear anything,' said Ridcully.

'I'm positive,' said Ponder, edging towards the door. 'I'll tell you what, sir, I'll just go and check.'

FOURTEEN

ALEPH-UMPTYPLEX

THE WIZARDS ARE NOT ONLY grappling with the apparent absurdities of 'quantum', their catch-all phrase for advanced physics and cosmology, but with the explosive philosophical/ mathematical concept of infinity. In their own way, they have rediscovered one of the great insights of nineteenth-century mathematics: that there can be many infinities, some of them bigger than others.

If this sounds ridiculous, it is. Nonetheless, there is an entirely natural sense in which it turns out to be true.

There are two important things to understand about infinity. Although the infinite is often compared with numbers like 1, 2, 3, infinity is not itself a number in any conventional sense. As Ponder Stibbons says, you can't get there from 1. The other is that, even within mathematics, there are many distinct notions that all bear the same label 'infinity'. If you mix up their meanings, all you'll get is nonsense.

And then – sorry, *three* important things – you have to appreciate that infinity is often a process, not a thing.

But – oh, *four* important things – mathematics has a habit of turning processes into things.

Oh, and – all right, *five* important things – one kind of infinity *is* a number, though a slightly unconventional one.

As well as the mathematics of infinity, the wizards are also contending with its physics. Is the Roundworld universe finite or infinite? Is it true that in any infinite universe, not only *can* anything happen, but everything must? Could there be an infinite universe consisting entirely of chairs . . . immobile, unchanging, wildly unexciting? The world of the infinite is paradoxical, or so it seems at first, but we shouldn't let the apparent paradoxes put us off. If we keep a clear head, we can steer our way through the paradoxes, and turn the infinite into a reliable thinking aid.

Philosophers generally distinguish two different 'flavours' of infinity, which they call 'actual' and 'potential'. Actual infinity is a *thing* that is infinitely big, and that's such a mouthful to swallow that until recently it was rather disreputable. The more respectable flavour is potential infinity, which arises whenever some process gives us the distinct impression that it could be continued for as long as we like. The most basic process of this kind is counting: 1, 2, 3, 4, 5 . . . Do we ever reach 'the biggest possible number' and then stop? Children often ask that question, and at first they think that the biggest number whose name they *know* must be the biggest number there *is*. So for a while they think that the biggest number is six, then they think it's a hundred, then they think it's a thousand. Shortly after, they realise that if you can count to a thousand, then a thousand and one is only a single step further.

In their 1949 book *Mathematics and the Imagination*, Edward Kasner and James Newman introduced the world to the googol – the digit 1 followed by a hundred zeros. Bear in mind that a billion has a mere nine zeros: 1000000000. A googol is

10000000000000000000000000000000000000000000000000
00000000000000000000000000000000000000000000000000

and it's so big we had to split it in two to fit the page. The name was invented by Kasner's nine-year-old nephew, and is the inspiration for the internet search engine Google™.

Even though a googol is very big, it is definitely not infinite. It is easy to write down a bigger number:

10000000000000000000000000000000000000000000000000000
00000000000000000000000000000000000000000000000000001

Just add 1. A more spectacular way to find a bigger number than a googol is to form a googolplex (name also courtesy of the nephew), which is 1 followed by a googol of zeros. Do not attempt to write this number down: the universe is too small unless you use sub-atomic-sized digits, and its lifetime is too short, let alone yours.

Even though a googolplex is extraordinarily big, it is a precisely defined number. There is nothing vague about it. And it is definitely not infinite (just add 1). It is, however, big enough for most purposes, including most numbers that turn up in astronomy. Kasner and Newman observe that 'as soon as people talk about large numbers, they run amuck. They seem to be under the impression that since zero equals nothing, they can add as many zeros to a number as they please with practically no serious consequences,' a sentence the Mustrum Ridcully himself might have uttered. As an example, they report that in the late 1940s a distinguished scientific publication announced that the number of snow crystals needed to start an ice age is a billion to the billionth power. 'This,' they tell us, 'is very startling and also very silly.' A billion to the billionth power is 1 followed by nine billion zeros. A sensible figure is around 1 followed by 30 zeros, which is fantastically smaller, though still bigger than Bill Gates's bank balance.

Whatever infinity may be, it's not a conventional 'counting' number. If the biggest number possible were, say, umpty-ump gazillion, then by the same token umpty-ump gazillion *and one* would be bigger still. And even if it were more complicated, so that (say) the biggest number possible were umpty-ump gazillion, two million, nine hundred and sixty-four thousand, seven hundred and fifty-eight . . . then what about umpty-ump gazillion, two million, nine hundred and sixty-four thousand, seven hundred and fifty-*nine*?

Given any number, you can always add one, and then you get a number that is (slightly, but distinguishably) *bigger*.

The counting process only stops if you run out of breath; it does not stop because you've run out of *numbers*. Though a near-immortal might perhaps run out of universe in which to write the numbers down, or time in which to utter them.

In short: there exist infinitely many numbers.

The wonderful thing about that statement is that it does not imply that there is some number called 'infinity', which is bigger than any of the others. Quite the reverse: the whole point is that there *isn't* a number that is bigger than any of the others. So although the process of counting can in principle go on for ever, the number you have reached at any particular stage is finite. 'Finite' means that you can count up to that number and then stop.

As the philosophers would say: counting is an instance of potential infinity. It is a process that can go on for ever (or at least, so it seems to our naive pattern-recognising brains) but never *gets to* 'for ever'.

The development of new mathematical ideas tends to follow a pattern. If mathematicians were building a house, they would start with the downstairs walls, hovering unsupported a foot or so above the damp-proof course . . . or where the damp-proof course ought to be. There would be no doors or windows, just holes of the right shape. By the time the second floor was added, the quality of the brickwork would have improved dramatically, the interior walls would be plastered, the doors and windows would all be in place, and the floor would be strong enough to walk on. The third floor would be vast, elaborate, fully carpeted, with pictures on the walls, huge quantities of furniture of impressive but inconsistent design, six types of wallpaper in every room . . . The attic, in contrast, would be sparse but elegant – minimalist design, nothing out of place, everything there for a reason. Then, and only then, would they go back to ground level, dig the foundations, fill them with concrete, stick in a damp-proof course, and extend the walls downwards until they met the foundations.

At the end of it all you'd have a house that would stand up. Along the way, it would have spent a lot of its existence looking wildly improbable. But the builders, in their excitement to push the walls skywards and fill the rooms with interior decor, would have been too busy to notice until the building inspectors rubbed their noses in the structural faults.

When new mathematical ideas first arise, no one understands them terribly well, which is only natural because they're *new*. And no one is going to make a great deal of effort to sort out all the logical refinements and make sense of those ideas unless they're convinced it's all going to be worthwhile. So the main thrust of research goes into developing those ideas and seeing if they lead anywhere interesting. 'Interesting', to a mathematician, mostly means 'can I see ways to push this stuff further?', but the acid test is 'what problems does it solve?' Only after getting a satisfactory answer to these questions do a few hardy and pedantic souls descend into the basement and sort out decent foundations.

So mathematicians were *using* infinity long before they had a clue what it was or how to handle it safely. In 500 BC Archimedes, the greatest of the Greek mathematicians and a serious contender for a place in the all-time top three, worked out the volume of a sphere by (conceptually) slicing it into infinitely many infinitely thin discs, like an ultra-thin sliced loaf, and hanging all the slices from a balance, to compare their total volume with that of a suitable shape whose volume he already knew. Once he'd worked out the answer by this astonishing method, he started again and found a logically acceptable way to prove he was right. But without all that faffing around with infinity, he wouldn't have known where to start and his logical proof wouldn't have got off the ground.

By the time of Leonhard Euler, an author so prolific that we might consider him to be the Terry Pratchett of eighteenth-century mathematics, many of the leading mathematicians were dabbling in 'infinite series' – the school child's nightmare of *a sum that never ends*. Here's one:

$$1 + 1/2 + 1/4 + 1/8 + 1/16 + 1/32 + \ldots$$

where the '. . .' means 'keep going'. Mathematicians have concluded that if this infinite sum adds up to anything sensible, then what it adds up to must be exactly two.* If you stop at any finite stage, though, what you reach is slightly less than two. But the amount by which it is less than two keeps shrinking. The sum sort of sneaks up on the correct answer, without actually getting there; but the amount by which it fails to get there can be made as small as you please, by adding up enough terms.

Remind you of anything? It looks suspiciously similar to one of Zeno/Xeno's paradoxes. This is how the arrow sneaks up on its victim, how Achilles sneaks up on the tortoise. It is how you can do infinitely many things in a finite time. Do the first thing; do the second thing one minute later; do the third thing half a minute after that; then the fourth thing a quarter of a minute after *that* . . . and so on. After two minutes, you've done infinitely many things.

The realisation that infinite sums *can* have a sensible meaning is only the start. It doesn't dispel all of the paradoxes. Mostly, it just

---

* To see why, double it: the result now is $2 + 1 + \frac{1}{2} + \frac{1}{4} + \frac{1}{8} + \frac{1}{16} + \ldots$, which is 2 more than the original sum. What number increases by 2 when you double it? There's only one such number, and it's 2.

sharpens them. Mathematicians worked out that some infinities are harmless, others are not.

The only problem left after that brilliant insight was: how do you tell? The answer is that if your concept of infinity does not lead to logical contradictions, then it's safe to use, but if it does, then it isn't. Your task is to *give* a sensible meaning to whatever 'infinity' intrigues you. You can't just assume that it automatically makes sense.

Throughout the eighteenth and early nineteenth centuries, mathematics developed many notions of 'infinity', all of them potential. In projective geometry, the 'point at infinity' was where two parallel lines met: the trick was to draw them in perspective, like railway lines heading off towards the horizon, in which case they appear to meet *on* the horizon. But if the trains are running on a plane, the horizon is infinitely far away and it isn't actually part of the plane at all – it's an optical illusion. So the point 'at' infinity is determined by the *process* of travelling along the train tracks indefinitely. The train never actually gets there. In algebraic geometry a circle ended up being *defined* as 'a conic section that passes through the two imaginary circular points at infinity', which sure puts a pair of compasses in their place.

There was an overall consensus among mathematicians, and it boiled down to this. Whenever you use the term 'infinity' you are really thinking about a process. If that process generates some well-determined *result*, by however convoluted an interpretation you wish, then that result gives meaning to your use of the word 'infinity', in that particular context.

Infinity is a context-dependent process. It is potential.

It couldn't stay that way.

David Hilbert was one of the top two mathematicians in the world at the end of the nineteenth century, and he was one of the great enthusiasts for a new approach to the infinite, in which – contrary to what we've just told you – infinity is treated as a thing, not as a process. The new approach was the brainchild of Georg Cantor, a German mathematician whose work led him into territory that was fraught with logical snares. The whole area was a confused mess for about a century (nothing new there, then). Eventually he decided to sort it out for good and all by burrowing downwards rather than building ever upwards, and putting in those previously non-existent foundations. He wasn't the only person doing this, but he was among the more radical ones. He succeeded in sorting out the area that drove him to these lengths, but only at the expense of causing considerable trouble elsewhere.

Many mathematicians detested Cantor's ideas, but Hilbert loved them, and defended them vigorously. 'No one,' he declaimed, 'shall expel us from the paradise that Cantor has created.' It is, to be sure, as much paradox as paradise. Hilbert explained some of the paradoxical properties of infinity à la Cantor in terms of a fictitious hotel, now known as Hilbert's Hotel.

Hilbert's Hotel has infinitely many rooms. They are numbered 1, 2, 3, 4 and so on indefinitely. It is an instance of actual infinity – every room exists *now*, they're not still building room umpty-ump gazillion and one. And when you arrive there, on Sunday morning, every room is occupied.

In a finite hotel, even with umpty-ump gazillion and one rooms, you're in trouble. No amount of moving people around can create an extra room. (To keep it simple, assume no sharing: each room has exactly one occupant, and health and safety regulations forbid more than that.)

In Hilbert's Hotel, however, there is always room for an extra guest. Not in room infinity, though, for there is no such room. In room one.

But what about the poor unfortunate in room one? He gets moved to room two. The person in room two is moved to room three. And so on. The person in room umpty-ump gazillion is moved to room umpty-ump gazillion and one. The person in room umpty-ump gazillion and one is moved to room umpty-ump gazillion and two.

The person in room n is moved to room $n+1$, for every number n.

In a finite hotel with umpty-ump gazillion and one rooms, this procedure hits a snag. There is no room umpty-ump gazillion and two into which to move its inhabitant. In Hilbert's Hotel, there is no end to the rooms, and everyone can move one place up. Once this move is completed, the hotel is once again full.

That's not all. On Monday, a coachload of 50 people arrives at the completely full Hilbert Hotel. No worries: the manager moves everybody up 50 places – room 1 to 51, room 2 to 52, and so on – which leaves rooms 1–50 vacant for the people off the coach.

On Tuesday, an Infinity Tours coach arrives containing infinitely many people, helpfully numbered A1, A2, A3, ... Surely there won't be room now? But there is. The existing guests are moved into the even-numbered rooms: room 1 moves to room 2, room 2 to room 4, room 3 to room 6, and so on. Then the odd-numbered rooms are free, and person A1 goes into room 1, A2 into room 3, A3 into room 5 . . . Nothing to it.

By Wednesday, the manager is really tearing his hair out, because *infinitely many* Infinity Tours coaches turn up. The coaches are labelled A, B, C, . . . from an infinitely long alphabet, and the people in them are A1, A2, A3, . . . , B1, B2, B3, . . . C1, C2, C3, . . . and so on. But the manager has a brainwave. In an infinitely large corner of the infinitely large hotel parking lot, he arranges all the new guests into an infinitely large square:

A1 A2 A3 A4 A5 . . .
B1 B2 B3 B4 B5 . . .
C1 C2 C3 C4 C5 . . .
D1 D2 D3 D4 D5 . . .
E1 E2 E3 E4 E5 . . .
   . . .

Then he rearranges them into a single infinitely long line, in the order

A1 - A2 B1 - A3 B2 C1 - A4 B3 C2 D1 - A5 B4 C3 D2 E1 ...

(To see the pattern, look along successive diagonals running from top right to lower left. We've inserted hyphens to separate these.) What most people would now do is move all the existing guests into the even-numbered rooms, and then fill up the odd rooms with new guests, in the order of the infinitely long line. That works, but there is a more elegant method, and the manager, being a mathematician, spots it immediately. He loads everybody back into a single Infinity Tours coach, filling the seats in the order of the infinitely long line. This reduces the problem to one that has already been solved.*

Hilbert's Hotel tells us to be careful when making assumptions about infinity. It may not behave like a traditional finite number. If you add one to infinity, it doesn't get bigger. If you multiply infinity by infinity, it *still* doesn't get bigger. Infinity is like that. In fact, it's easy to conclude that *any* sum involving infinity works out as infinity, because you can't get anything bigger than infinity.

That's what everybody thought, which is fair enough if the only infinities you've ever encountered are potential ones, approached as a sequence of finite steps, but in principle going on for as long as you wish. But in the 1880s Cantor was thinking about actual

---

* If you've never encountered the mathematical joke, here it is. Problem 1: a kettle is hanging on a peg. Describe the sequence of events needed to make a pot of tea. Answer: take the kettle off the peg, put it in the sink, turn on the tap, wait till the kettle fills with water, turn the tap off . . . and so on. Problem 2: a kettle is sitting in the sink. Describe the sequence of events needed to make a pot of tea. Answer: *not* 'turn on the tap, wait till the kettle fills with water, turn the tap off . . . and so on'. Instead: take the kettle out of the sink and *hang it on the peg*; then proceed as before. This reduces the problem to one that has already been solved. (Of course the first step puts it back in the sink – that's why it's a joke.)

infinities, and he opened up a veritable Pandora's box of ever-larger infinities. He called them *transfinite numbers*, and he stumbled across them when he was working in a hallowed, traditional area of analysis. It was really hard, technical stuff, and it led him into previously uncharted byways. Musing deeply on the nature of these things, Cantor became diverted from his work in his entirely respectable area of analysis, and started thinking about something much more difficult.

Counting.

The usual way that we introduce numbers is by teaching children to count. They learn that numbers are 'things you use for counting'. For instance, 'seven' is where you get to if you start counting with 'one' for Sunday and stop on Saturday. So the number of days in the week is seven. But what manner of beast is seven? A word? No, because you could use the symbol 7 instead. A symbol? But then, there's the word . . . anyway, in Japanese, the symbol for 7 is different. So what *is* seven? It's easy to say what seven days, or seven sheep, or seven colours of the spectrum are . . . but what about the *number* itself? You never encounter a naked 'seven', it always seems to be attached to some collection of things.

Cantor decided to make a virtue of necessity, and declared that a number was something associated with a set, or collection, of things. You can put together a set from any collection of things whatsoever. Intuitively, the number you get by counting tells you how many things belong to that set. The set of days of the week determines the number 'seven'. The wonderful feature of Cantor's approach is this: you can decide whether any other set has seven members *without* counting anything. To do this, you just have to try to match the members of the sets, so that each member of one set is matched to precisely one of the other. If, for instance, the second set is the set of colours of the spectrum, then you might match the sets like this:

Sunday   Red
Monday   Orange

Tuesday	Yellow
Wednesday	Green
Thursday	Blue
Friday	Violet*
Saturday	Octarine

The order in which the items are listed does not matter. But you're not allowed to match Tuesday with both Violet and Green, or Green with both Tuesday and Sunday, in the same matching. Or to miss any members of the sets out.

In contrast, if you try to match the days of the week with the elephants that support the Disc, you run into trouble:

Sunday	Berilia
Monday	Tubul
Tuesday	Great T'Phon
Wednesday	Jerakeen
Thursday	?

More precisely, you run out of elephants. Even the legendary fifth elephant fails to take you past Thursday.

Why the difference? Well, there are seven days in the week, and seven colours of the spectrum, so you can match those sets. But there are only four (perhaps once five) elephants, and you can't match four or five with seven.

The deep philosophical point here is that you don't need to know about the *numbers* four, five or seven, to discover that there's no way to match the sets up. Talking about the numbers amounts to being wise after the event. Matching is logically primary

* Yes, traditionally 'Indigo' goes here, but that's silly – Indigo is just another shade of blue. You could equally well insert 'Turquoise' between Green and Blue. Indigo was just included because seven is more mystical than six. Rewriting history, we find that we have left a place for Octarine, the Discworld's eighth colour. Well, seventh, actually. Septarine, anyone?

to counting.* But now, all sets that match each other can be assigned a common symbol, or 'cardinal', which effectively is the corresponding number. The cardinal of the set of days of the week is the symbol 7, for instance, and the same symbol applies to any set that matches the days of the week. So we can base our concept of number on the simpler one of matching.

So far, then, nothing new. But 'matching' makes sense for infinite sets, not just finite ones. You can match the even numbers with all numbers:

2   1
4   2
6   3
8   4
10   5

. . .

and so on. Matchings like this explain the goings-on in Hilbert's Hotel. That's where Hilbert got the idea (roof before foundations, remember).

What is the cardinal of the set of all whole numbers (and hence of any set that can be matched to it)? The traditional name is 'infinity'. Cantor, being cautious, preferred something with fewer mental associations, and in 1883 he named it 'aleph', the first letter of the Hebrew alphabet. And he put a small zero underneath it, for reasons that will shortly transpire: aleph-zero.

He knew what he was starting: 'I am well aware that by adopting

---

* This is why, even today when the lustre of 'the new mathematics' has all but worn to dust, small children in mathematics classes spend hours drawing squiggly lines between circles containing pictures of cats to circles containing pictures of flowers, busily 'matching' the two sets. Neither the children nor their teachers have the foggiest idea *why* they are doing this. In fact they're doing it because, decades ago, a bunch of demented educators couldn't understand that just because something is *logically* prior to another, it may not be sensible to teach them in that order. Real mathematicians, who knew that you always put the roof on the house *before* you dug the foundation trench, looked on in bemused horror.

such a procedure I am putting myself in opposition to widespread views regarding infinity in mathematics and to current opinions on the nature of number.' He got what he expected: a lot of hostility, especially from Leopold Kronecker. 'God created the integers: all else is the work of Man,' Kronecker declared.

Nowadays, most of us think that Man created the integers too.

Why introduce a new symbol (and Hebrew at that?). If there had been only one infinity in Cantor's sense, he might as well have named it 'infinity' like everyone else, and used the traditional symbol of a figure 8 lying on its side. But he quickly saw that from his point of view, there might well be other infinities, and he was reserving the right to name those aleph-one, aleph-two, aleph-three, and so on.

How can there be *other* infinities? This was the big unexpected consequence of that simple, childish idea of matching. To describe how it comes about, we need some way to talk about really big numbers. Finite ones and infinite ones. To lull you into the belief that everything is warm and friendly, we'll introduce a simple convention.

If 'umpty' is any number, of whatever size, then 'umptyplex' will mean 10^{umpty}, which is 1 followed by umpty zeros. So 2plex is 100, a hundred; 6plex is 1000000, a million; 9plex is a billion. When umpty = 100 we get a googol, so googol = 100plex. A googolplex is therefore also describable as 100plexplex.

In Cantorian mode, we idly start to muse about infinityplex. But let's be precise: what about aleph-zeroplex? What is $10^{aleph\text{-}zero}$?

Remarkably, it has an entirely sensible meaning. It is the cardinal of the set of all real numbers – all numbers that can be represented as an infinitely long decimal. Recall the Ephebian philosopher Pthagonal, who is recorded as saying, 'The diameter divides into the circumference . . . It ought to be three times. But does it? No. Three point one four and lots of other figures. There's no end to the buggers.' This, of course, is a reference to the most famous real number, one that really does need infinitely many decimal places to capture it exactly: π ('pi'). To one decimal place, π is 3.1. To two places, it is 3.14. To three places, it is 3.141. And so on, ad infinitum.

There are plenty of real numbers other than π. How big is the phase space of all real numbers?

Think about the bit after the decimal point. If we work to one decimal place, there are 10 possibilities: any of the digits 0, 1, 2, . . . , 9. If we work to two decimal places, there are 100 possibilities: 00 up to 99. If we work to three decimal places, there are 1000 possibilities: 000 up to 999.

The pattern is clear. If we work to umpty decimal places, there are 10^{umpty} possibilities. That is, umptyplex.

If the decimal places go on 'for ever', we first must ask 'what kind of for ever?' And the answer is 'Cantor's aleph-zero', because there is a first decimal place, a second, a third . . . the places match the whole numbers. So if we set 'umpty' equal to 'aleph-zero', we find that the cardinal of the set of all real numbers (ignoring anything before the decimal point) is aleph-zeroplex. The same is true, for slightly more complicated reasons, if we *include* the bit before the decimal point.[*]

All very well, but presumably aleph-zeroplex is going to turn out to be aleph-zero in heavy disguise, since all infinities surely must be equal? No. They're not. Cantor proved that you can't match the real numbers with the whole numbers. So aleph-zeroplex is a bigger infinity than aleph-zero.

He went further. Much further. He proved[†] that if umpty is any infinite cardinal, the umptyplex is a bigger one. So aleph-zeroplexplex is

---

[*] Briefly: since the bit before the decimal point is a whole number, taking that into account multiplies the answer by aleph-zero. Now aleph-zero $\times$ aleph-zeroplex is less than or equal to aleph-zeroplex $\times$ aleph-zeroplex, which is (2 $\times$ aleph-zero)plex, which is aleph-zeroplex. OK?

[†] The proof isn't hard, but it's sophisticated. If you want to see it, consult a textbook on the foundations of mathematics.

bigger still, and aleph-zeroplexplexplex is bigger than that, and . . .

There is no end to the list of Cantorian infinities. There is no 'hyperinfinity' that is bigger than all other infinities.

The idea of infinity as 'the biggest possible number' is taking some hard knocks here. And this is the *sensible* way to set up infinite arithmetic.

If you start with any infinite cardinal aleph-umpty, then aleph-umptyplex is bigger. It is natural to suppose that what you get must be aleph-(umpty+1), a statement dubbed the Generalised Continuum Hypothesis. In 1963 Paul Cohen (no known relation either to Jack or the Barbarian) proved that . . . well, it depends. In some versions of set theory it's true, in others it's false.

The foundations of mathematics are like that, which is why it's best to construct the house first and put the foundations in later. That way, if you don't like them, you can take them out again and put something else in instead. Without disturbing the house.

This, then, is Cantor's Paradise: an entirely new number system of alephs, of infinities beyond measure, never-ending – in a very strong sense of 'never'. It arises entirely naturally from one simple principle: that the technique of 'matching' is all you need to set up the logical foundations of arithmetic. Most working mathematicians now agree with Hilbert, and Cantor's initially astonishing ideas have been woven into the very fabric of mathematics.

The wizards don't just have the mathematics of infinity to contend with. They are also getting tangled up in the physics. Here, entirely new questions about the infinite arise. Is the universe finite or infinite? What *kind* of finite or infinite? And what about all those parallel universes that the cosmologists and quantum theorists are always talking about? Even if each universe is finite, could there be infinitely many parallel ones?

According to current cosmology, what we normally think of as the universe is finite. It started as a single point in the Big Bang, and then expanded at a finite rate for about 13 billion years, so it has to be finite. Of course, it could be infinitely finely divisible, with no lower limit to the sizes of things, just like the mathematician's line or plane – but quantum-mechanically speaking there is a definite graininess down at the Planck length, so the universe has a very large but finite number of possible quantum states.

The 'many worlds' version of quantum theory was invented by the physicist Hugh Everett as a way to link the quantum view of the world to our everyday 'sensible' view. It contends that whenever a choice can be made – for example, whether an electron spin is up or down, or a cat is alive or dead – the universe does not simply make a choice and abandon all the alternatives. That's what it looks like to us, but really the universe makes *all possible* choices. Innumerable 'alternative' or 'parallel' worlds branch off from the one that we perceive. In those worlds, things happen that did not happen here. In one of them, Adolf Hitler won the Second World War. In another, you ate one extra olive at dinner last night.

Narratively speaking, the many worlds description of the quantum realm is a delight. No author in search of impressive scientific gobbledegook that can justify hurling characters into alternative storylines – we plead guilty – can possibly resist.

The trouble is that, as science, the many-worlds interpretation is rather overrated. Certainly, the usual way that it is described is misleading. In fact, rather too much of the physics of multiple universes is usually explained in a misleading way. This is a pity, because it trivialises a profound and beautiful set of ideas. The suggestion that there exists a real universe, somehow adjacent to ours, in which Hitler defeated the Allies, is a big turn-off for a lot of people. It sounds too absurd even to be worth considering. 'If that's what modern physics is about, I'd prefer my tax dollars to go towards something useful, like reflexology.'

The science of 'the' multiverse – there are numerous alternatives, which is only appropriate – is fascinating. Some of it is even useful.

And some – not necessarily the useful bit – might even be true. Though not, we will try to convince you, the bit about Hitler.

It all started with the discovery that quantum behaviour can be represented mathematically as a Big Sum. What actually happens is the sum of all of the things that might have happened. Richard Feynman explained this with his usual extreme clarity in his book *QED* (**Q**uantum **E**lectro **D**ynamics, not Euclid). Imagine a photon, a particle of light, bouncing off a mirror. You can work out the path that the photon follows by 'adding up' all possible paths that it might have taken. What you really add is the levels of brightness, the light intensities, not the paths. A path is a concentrated strip of brightness, and here that strip hits the mirror and bounces back at the same angle.

This 'sum-over-histories' technique is a direct mathematical consequence of the rules of quantum mechanics, and there's nothing objectionable or even terribly surprising about it. It works because all of the 'wrong' paths interfere with each other, and between them they contribute virtually nothing to the overall sum. All that survives, as the totals come in, is the 'right' path. You can take this unobjectionable mathematical fact and dress it up with a physical interpretation. Namely: light *really* takes all possible paths, but what we observe is the sum, so we just see the one path in which the light 'ray' hits the mirror and bounces off again at the same angle.

That interpretation is also not terribly objectionable, philosophically speaking, but it verges into territory that is. Physicists have a habit of taking mathematical descriptions literally – not just the conclusions, but the steps employed to get them. They call this 'thinking physically', but actually it's the reverse: it amounts to projecting mathematical features on to the real world – 'reifying' abstractions, endowing them with reality.

We're not saying it doesn't *work* – often it does. But reification tends to make physicists bad philosophers, because they forget they're doing it.

One problem with 'thinking physically' is that there are sometimes several mathematically equivalent ways to describe something – different ways to say exactly the same thing in mathematical language. If one of them is true, they all are. But, their natural physical interpretations can be inconsistent.

A good example arises in classical (non-quantum) mechanics. A moving particle can be described using (one of) Newton's laws of motion: the particle's acceleration is proportional to the forces that act on it. Alternatively, the motion can be described in terms of a 'variational principle': associated with each possible particle path there is a quantity called the 'action'. The actual path that the particle follows is the one that makes the action as small as possible.

The logical equivalence of Newton's laws and the principle of least action is a mathematical theorem. You cannot accept one without accepting the other, on a mathematical level. Don't worry what 'action' is. It doesn't matter here. What matters is the difference between the natural interpretations of these two logically identical descriptions.

Newton's laws of motion are local rules. What the particle does next, here and now, is entirely determined by the forces that act on it, here and now. No foresight or intelligence is needed; just keep on obeying the local rules.

The principle of least action has a different style: it is global. It tells us that in order to move from A to B, the particle must somehow contemplate the totality of all possible paths between those points. It must work out the action associated with each path, and find whichever one of them has the smallest action. This 'computation' is non-local, because it involves the entire path(s), and in some sense it has to be carried out *before the particle knows where to go*. So in this natural interpretation of the mathematics, the particle appears to be endowed with miraculous foresight and the ability to choose, a rudimentary kind of intelligence.

So which is it? A mindless lump of matter which obeys the local rules as it goes along? Or a quasi-intelligent entity with vast computational powers, which has the foresight to choose, among all the possible paths that it could have taken, precisely the unique one that minimises the action?

We know which interpretation we'd choose.

Interestingly, the principle of least action is a mechanical analogue of Feynman's sum-over-histories method in optics. The two really are extremely close. Yes, you can formulate the mathematics of quantum mechanics in a way that seems to imply that light follows all possible paths and adds them up. But you are not obliged to buy that description as the real physics of the real world, even if the mathematics works.

The many-worlds enthusiasts do buy that description: in fact, they take it much further. Not the history of a single photon bouncing off a mirror, but the history of the entire universe. That, too, is a sum of all possibilities – using the universe's quantum wave function in place of the light intensity due to the photon – so by the same token, we can interpret the mathematics in a similarly dramatic way. Namely: the universe really does do all possible things. What we observe is what happens when you add all those possibilities up.

Of course there's also a less dramatic interpretation: the universe trundles along obeying the local laws of quantum mechanics, and does exactly *one* thing . . . which just happens, for purely mathematical reasons, to equal the sum of all the things that it might have done.

Which interpretation do you buy?

Mathematically, if one is 'right' then so is the other. Physically, though, they carry very different implications about how the world works. Our point is that, as for the classical particle, their mathematical equivalence does not require you to accept their physical truth as descriptions of reality. Any more than the equivalence of Newton's laws with the principle of least action obliges you to believe in intelligent particles that can predict the future.

The many-worlds interpretation of quantum mechanics, then, is resting on dodgy ground even though its mathematical foundations are impeccable. But the usual presentation of that interpretation goes further, by adding a hefty dose of narrativium. This is precisely what appeals to SF authors, but it's a pity that it stretches the interpretation well past breaking-point.

What we are usually told is this. At every instant of time, whenever a choice has to be made, the universe splits into a series of 'parallel worlds' in which each of the choices happens. Yes, in this world you got up, had cornflakes for breakfast, and walked to work. But somewhere 'out there' in the vastness of the multiverse, there is another universe in which you had kippers for breakfast, which made you leave the house a minute later, so that when you walked across the road you had an argument with a bus, and lost, fatally.

What's wrong here is not, strangely enough, the contention that this world is 'really' a sum of many others. Perhaps it is, on a quantum level of description. Why not? But it is wrong to describe those alternative worlds in human terms, as scenarios where everything follows a narrative that makes sense to the human mind. As worlds where 'bus' or 'kipper' have any meaning at all. And it is even less justifiable to pretend that every single one of those parallel worlds is a minor variation on this one, in which some human-level choice happens differently.

If those parallel worlds exist at all, they are described by changing various components of a quantum wave function whose complexity is beyond human comprehension. The results need not resemble humanly comprehensible scenarios. Just as the sound of a clarinet can be decomposed into pure tones, but most combinations of those tones do not correspond to any clarinet.

The natural components of the human world are buses and

kippers. The natural components of the quantum wave function of the world are not the quantum wave functions of buses and kippers. They are altogether different, and they carve up reality in a different way. They flip electron spins, rotate polarisations, shift quantum phases.

They do not turn cornflakes into kippers.

It's like taking a story and making random changes to the letters, shifting words around, probably changing the instructions that the printer uses to make the letters, so that they correspond to no alphabet known to humanity. Instead of starting with the Ankh-Morpork national anthem and getting the Hedgehog Song, you just get a meaningless jumble. Which is perhaps as well.

According to Max Tegmark, writing in the May 2003 issue of *Scientific American*, physicists currently recognise four distinct levels of parallel universes. At the first level, some distant region of the universe replicates, almost exactly, what is going on in our own region. The second level involves more or less isolated 'bubbles', baby universes, in which various attributes of the physical laws, such as the speed of light, are different, though the basic laws are the same. The third level is Everett's many-worlds quantum parallelism. The fourth includes universes with radically different physical laws – not mere variations on the theme of our own universe, but totally distinct systems described by every conceivable mathematical structure.

Tegmark makes a heroic attempt to convince us that all of these levels really do exist – that they make testable predictions, are scientifically falsifiable if wrong, and so on. He even manages to reinterpret Occam's razor, the philosophical principle that explanations should be kept as simple as possible, to support his view.

All of this, speculative as it may seem, is good frontier cosmology and physics. It's exactly the kind of theorising that a *Science of Discworld* book ought to discuss: imaginative, mind-boggling, cutting-edge. We've come to the reluctant conclusion, though, that the arguments have serious flaws. This is a pity, because the concept of parallel worlds is dripping with enough narrativium to make any SF author out-salivate Pavlov's dogs.

We'll summarise Tegmark's main points, describe some of the evidence that he cites in their favour, offer a few criticisms, and leave you to form your own opinions.

Level 1 parallel worlds arise if – because – space is infinite. Not so far back we told you it is finite, because the Big Bang happened a finite time ago so it's not had time to expand to an infinite extent.* Apparently, though, data on the cosmic microwave background do not support a finite universe. Even though a very large finite one would generate the same data.

'Is there a copy of you reading this article?' Tegmark asks. Assuming the universe is infinite, he tells us that 'even the most unlikely events must take place somewhere'. A copy of you is likelier than many, so it must happen. Where? A straightforward calculation indicates that 'you have a twin in a galaxy about 10 to the power 10^{28} metres from here'. Not 10^{28} metres, which is already 25 times the size of the currently observable universe, but 1 followed by 10^{28} zeros. Not only that: a complete copy of (the observable part of) our universe should exist about 10 to the power 10^{118} metres away. And beyond that . . .

We need a good way to talk about very big numbers. Symbols like 10^{118} are too formal. Writing out all the zeros is pointless, and usually impossible. The universe is big, and the multiverse is substantially bigger. Putting numbers to *how* big is not entirely straightforward, and finding something that can also be typeset is even harder.

---

* Curiously, it could expand to infinity in a finite time if it accelerated sufficiently rapidly. Expand by one light-year after one minute, by another light-year after half a minute, by another after a quarter of a minute . . . do a Zeno, and after two minutes, you have an infinite universe. But it's not expanding that fast, and no one thinks it did so in the past, either.

Fortunately, we've already solved that problem with our earlier convention: if 'umpty' is any number, then 'umptyplex' will mean 10^{umpty}, which is 1 followed by umpty zeros.

When umpty = 118 we get 118plex, which is roughly the number of protons in the universe. When umpty is 118plex we get 118plexplex, which is the number that Tegmark is asking us to think about, 10 to the power 10 to the power 118. Those numbers arise because a 'Hubble volume' of space – one the size of the observable universe – has a large but finite number of possible quantum states.

The quantum world is grainy, with a lower limit to how far space and time can be divided. So a sufficiently large region of space will contain such a vast number of Hubble volumes that every one of those quantum states can be accommodated. Specifically, a Hubble volume contains 118plex protons. Each has two possible quantum states. That means there are 2 to the power 118plex possible configurations of quantum states of protons. One of the useful rules in this type of mega-arithmetic is that the 'lowest' number in the plexified stack – here 2 – can be changed to something more convenient, such as 10, without greatly affecting the top number. So, in round numbers, a region 118plexplex metres across can contain one copy of each Hubble volume.

Level 2 worlds arise on the assumption that spacetime is a kind of foam, in which each bubble constitutes a universe. The main reason for believing this is 'inflation', a theory that explains why our universe is relativistically flat. In a period of inflation, space rapidly stretches, and it can stretch so far that the two ends of the stretched bit become independent of each other because light can't get from one to the other fast enough to connect them causally. So spacetime ends up as a foam, and each bubble probably has its own variant of the laws of physics – with the same basic mathematical form, but different constants.

Level 3 parallel worlds are those that appear in the many-worlds interpretation of quantum mechanics, which we've already tackled.

Everything described so far pales into insignificance when we come to level 4. Here, the various universes involved can have radically different laws of physics from each other. All conceivable mathematical structures, Tegmark tells us, exist here:

> How about a universe that obeys the laws of classical physics, with no quantum effects? How about time that comes in discrete steps, as for computers, instead of being continuous? How about a universe that is simply an empty dodecahedron? In the level IV multiverse, all these alternative realities actually exist.

But do they?

In science, you get evidence from observations or from experiments.

Direct observational tests of Tegmark's hypothesis are completely out of the question, at least until some remarkable spacefaring technology comes into being. The observable universe extends no more than 27plex metres from the Earth. An object (even the size of our visible universe) that is 118plexplex metres away cannot be observed now, and no conceivable improvement on technology can get round that. It would be easier for a bacterium to observe the entire known universe than for a human to observe an object 118plexplex metres away.

We are sympathetic to the argument that the impossibility of direct experimental tests does not make the theory unscientific. There is no *direct* way to test the previous existence of dinosaurs, or the timing (or occurrence) of the Big Bang. We infer these things from indirect evidence. So what indirect evidence is there for infinite space and distant copies of our own world?

Space is infinite, Tegmark says, because the cosmic microwave background tells us so. If space were finite, then traces of that finitude would show up in the statistical properties of the cosmic background and the various frequencies of radiation that make it up.

This is a curious argument. Only a year or so ago, some mathematicians used certain statistical features of the cosmic microwave background to deduce that not only is the universe finite, but that it is shaped a bit like a football.* There is a paucity of very long-wavelength radiation, and the best reason for not finding it is that the universe is too small to accommodate such wavelengths. Just as a guitar string a metre long cannot support a vibration with a wavelength of 100 metres – there isn't room to fit the wave into the available space.

The main other item of evidence is of a very different nature – not an observation as such, but an observation about how we interpret observations. Cosmologists who analyse the microwave background to work out the shape and size of the universe habitually report their findings in the form 'there is a probability of one in a thousand that such and such a shape and size could be consistent with the data'. Meaning that with 99.9 per cent probability we rule out that size and shape. Tegmark tells us that one way to interpret this is that at most one Hubble volume in a thousand, of that size and shape, would exhibit the observed data. 'The lesson is that the multiverse theory can be tested and falsified even when we cannot see the other universes. The key is to predict what the ensemble of parallel universes is and to specify a probability distribution over that ensemble.'

This is a remarkable argument. Fatally, it confuses actual Hubble volumes with potential ones. For example, if the size and shape under consideration is 'a football about 27plex metres across' – a fair guess for our own Hubble volume – then the 'one in a thousand' probability is a calculation based on a potential array of one thousand footballs of that size. These are not part of a single infinite universe: they are distinct conceptual 'points' in a phase space of big

---

* Actually a more sophisticated gadget called the Poincaré dodecahedral space, a slightly weird shape invented more than a century ago to show that topology is not as simple as we'd like it to be. But people understand 'football'.

footballs. If you lived in such a football and made such observations, then you'd expect to get the observed data on about one occasion in a thousand.

There is nothing in this statement that compels us to infer the actual existence of those thousand footballs – let alone to embed the lot in a single, bigger space, which is what we are being asked to do. In effect, Tegmark is asking us to accept a general principle: that whenever you have a phase space (statisticians would say a sample space) with a well-defined probability distribution, then everything in that phase space must be real.

This is plain wrong.

A simple example shows why. Suppose that you toss a coin a hundred times. You get a series of tosses something like HHTTTHH . . . TTHH. The phase space of all possible such tosses contains precisely 2^{100} such sequences. Assuming the coin is fair, there is a sensible way to assign a probability to each such sequence – namely the chance of getting it is one in 2^{100}. And you can test that 'distribution' of probabilities in various indirect ways. For instance, you can carry out a million experiments, each yielding a series of 100 tosses, and count what proportion has 50 heads and 50 tails, or 49 heads and 51 tails, whatever. Such an experiment is entirely feasible.

If Tegmark's principle is right, it now tells us that the entire phase space of coin-tossing sequences *really does exist*. Not as a mathematical concept, but as physical reality.

However, coins do not toss themselves. Someone has to toss them.

If you could toss 100 coins every second, it would take about 24plex years to generate 2^{100} experiments. That is roughly 100 trillion times the age of the universe. Coins have been in existence for only a few thousand years. The phase space of all sequences of 100 coin tosses is *not* real. It exists only as potential.

Since Tegmark's principle doesn't work for coins, it makes no sense to suppose that it works for universes.

The evidence advanced in favour of level 4 parallel worlds is even thinner. It amounts to a mystical appeal to Eugene Wigner's famous remark about 'the unusual effectiveness of mathematics' as a description of physical reality. In effect, Tegmark tells us that if we can imagine something, then it has to exist.

We can imagine a purple hippopotamus riding a bicycle along the edge of the Milky Way while singing Monteverdi. It would be lovely if that meant it had to exist, but at some point a reality check is in order.

We don't want to leave you with the impression that we enjoy pouring cold water over every imaginative attempt to convey a feeling for some of the remarkable concepts of modern cosmology and physics. So we'll end with a very recent addition to the stable of parallel worlds, which has quite a few things going for it. Perhaps unsurprisingly, the main thing *not* currently going for it is a shred of experimental evidence.

The new theory on the block is string theory. It provides a philosophically sensible answer to the age-old question: why are we here? And it does so by invoking gigantic numbers of parallel universes.

It is just much more careful how it handles them.

Our source is an article, 'The String Theory Landscape' by Raphael Bousso and Joseph Polchinski, in the September 2004 issue of *Scientific American* – a special issue on the theme of Albert Einstein.

If there is a single problem that occupies the core of modern physics, it is that of unifying quantum mechanics with relativity. This search for a 'theory of everything' is needed because although both of those theories are extraordinarily successful in helping us to understand and predict various aspects of the natural world, they are not totally consistent with each other. Finding a consistent, unified theory is hard, and we don't yet have one. But there's one mathematically attractive attempt, string theory, which is conceptually appealing even though there's no observational evidence for it.

String theory holds that what we usually consider to be individual points of spacetime, dimensionless dots with no interesting structure of their own, are actually very, very tiny multidimensional surfaces with complicated shapes. The standard analogy is a garden hose. Seen from some way off, a hose looks like a line, which is a one-dimensional space – the dimension being distance along the hose. Look more closely, though, and you see that the hose has two extra dimensions, at right angles to that line, and that its shape in those directions is a circular band.

Maybe our own universe is a bit like that hosepipe. Unless we look very closely, all we see is three dimensions of space plus one of time – relativity. An awful lot of physics is observed in those dimensions alone, so phenomena of that type have a nice four-dimensional description – relativity again. But other things might happen along extra 'hidden' dimensions, like the thickness of the hose. For instance, suppose that at each point of the apparent four-dimensional space-time, what seems to be a point is actually a tiny circle, sticking out at right angles to spacetime itself. That circle could vibrate. If so, then it would resemble the quantum description of a particle. Particles have various 'quantum numbers' such as spin. These numbers occur as whole number multiples of some basic amount. So do vibrations of a circle: either one wave fits into the circle, or two, or three . . . but not two and a quarter, say.

This is why it's called 'string theory'. Each point of spacetime is replaced by a tiny loop of string.

In order to reconstruct something that agrees with quantum the-ory, however, we can't actually use a circular string. There are too many distinct quantum numbers, and plenty of other problems that have to be overcome. The suggestion is that instead of a circle, we have to use a more complicated, higher-dimensional shape, known as a 'brane'.* Think of this as a surface, only more so. There are many

---

* Derived from a pun: *m*-brane for 'membrane'. Opening up jokes about no-branes and *p*-branes. Oh well.

distinct topological types of surface: a sphere, a doughnut, two doughnuts joined together, three doughnuts . . . and in more dimensions than two, there are more exotic possibilities.

Particles correspond to tiny closed strings that loop around the brane. There are lots of different ways to loop a string round a doughnut – once through the hole, twice, three times . . . The physical laws depend on the shape of the brane and the paths followed by these loops.

The current favourite brane has six dimensions, making ten in all. The extra dimensions are thought to be curled up very tightly, smaller than the Planck length, which is the size at which the universe becomes grainy. It is virtually impossible to observe anything that small, because the graininess blurs everything and the fine detail cannot be seen. So there's no hope of observing any extra dimensions directly. However, there are several ways to infer their presence indirectly. In fact, the recently discovered acceleration in the rate of expansion of the universe can be explained in that manner. Of course, this explanation may not be correct: we need more evidence.

The ideas here change almost by the day, so we don't have to commit ourselves to the currently favoured six-dimensional set-up. We can contemplate any number of different branes and differently arranged loops. Each choice – call it a loopy brane – has a particular energy, related to the shape of the brane, how tightly it is curled up, and how tightly the loops wind round it. This energy is the 'vacuum energy' of the associated physical theory. In quantum mechanics, a vacuum is a seething mass of particles and antiparticles coming into existence for a brief instant before they collide and annihilate each other again. The vacuum energy measures how violently they seethe. We can use the vacuum energy to infer which loopy brane corresponds to our own universe, whose vacuum energy is extraordinarily small. Until recently it was thought to be zero, but it's now thought to be about 1/120plex units, where a unit is one

Planck mass per cubic Planck length, which is a googol grammes per cubic metre.

We now encounter a cosmic 'three bears' story. Macho Daddy Bear prefers a vacuum energy larger than +1/118plex units, but such a spacetime would be subject to local expansions far more energetic than a supernova. Wimpy Mummy Bear prefers a vacuum energy smaller than −1/120plex units (note the minus sign), but then space-time contracts in a cosmic crunch and disappears. Baby Bear and Goldilocks like their vacuum energy to be 'just right': somewhere in the incredibly tiny range between +1/118plex and −1/120plex units. That is the Goldilocks zone in which life as we know it might possibly exist.

It is no coincidence that we inhabit a universe whose vacuum energy lies in the Goldilocks zone, because we *are* life as we know it. If we lived in any other kind of universe, we would be life as we don't know it. Not impossible, but not us.

This is our old friend the anthropic principle, employed in an entirely sensible way to relate the way we function to the kind of universe that we need to function in. The deep question here is not 'why do we live in a universe like that?', but 'why does there *exist* a universe like that, for us to live in?' This is the vexed issue of cosmological fine-tuning, and the improbability of a random universe hitting just the right numbers is often used to prove that something − they always say 'We don't know, could be an alien,' but what they're all thinking is: 'God' − must have set our universe up to be just right for us.

The string theorists are made of sterner stuff, and they have a more sensible answer.

In 2000 Bousso and Polchinski combined string theory with an earlier idea of Steven Weinberg to explain why we shouldn't be surprised that a universe with the right level of vacuum energy exists. Their basic idea is that the phase space of possible universes is absolutely gigantic. It is bigger than, say, 500plex. Those 500plex universes distribute their vacuum energies densely in the range -1 to +1 units. The resulting numbers are much more closely packed than the 1/118plex units that determine the scale of the 'acceptable' range of vacuum energies for life as we know it. Although only a very tiny proportion of those 500plex universes fall inside that range, there are so many of them that that a tiny proportion is still absolutely gigantic – here, around 382plex. So a whacking great 382plex universes, from a phase space of 500plex loopy branes, are capable of supporting our kind of life.

However, that's still a very small proportion. If you pick a loopy brane at random, the odds are overwhelmingly great that it won't fall inside the Goldilocks range.

Not a problem. The string theorists have an answer to that. If you wait long enough, such a universe will necessarily come into being. In fact, *all* universes in the phase space of loopy branes will eventually become the 'real' universe. And when the real universe's loopy brane gets into the Goldilocks range, the inhabitants of that universe will not know about all that waiting. Their sense of time will start from the instant when that particular loopy brane first occurred.

String theory not only tells us that we're here because we're here – it explains why a suitable 'here' must exist.

The reason why all of those 500plex or so universes can legitimately be considered 'real' in string theory stems from two features of that theory. The first is a systematic way to describe all the possible loopy branes that *might* occur. The second invokes a bit of quantum to explain why, in the long run, they *will* occur. Briefly: the phase space of loopy branes can be represented as an 'energy landscape', which we'll name the *branescape*. Each position in the landscape corresponds to one possible choice of loopy brane; the height at that point corresponds to the associated vacuum energy.

Peaks of the branescape represent loopy branes with high vacuum energy, valleys represent loopy branes with low vacuum energy. Stable loopy branes lie in the valleys. Universes whose hidden dimensions look like those particular loopy branes are themselves stable . . . so these are the ones that *can* exist, physically, for more than a split second.

In hilly districts of the world, the landscape is rugged, meaning that it has a lot of peaks and valleys. They get closer together than elsewhere, but they are still generally isolated from each other. The branescape is very rugged indeed, and it has a huge number of valleys. But all of the valleys' vacuum energies have to fit inside the range from -1 units to +1 units. With so many numbers to pack in, they get squashed very close together.

In order for a universe to support life as we know it, the vacuum energy has to lie in the Goldilocks zone where everything is just right. And there are so many loopy branes that a huge number of them must have vacuum energies that fall inside it.

Vastly more will fall outside that range, but never mind.

The theory has one major advantage: it explains why our universe has such a small vacuum energy, without requiring it to be zero – which, we now know, it isn't.

The upshot of all the maths, then, is that every stable universe sits in some valley of the branescape, and an awful lot of them (though a tiny proportion of the whole) lie in the Goldilocks range. But all of those universes are potential, not actual. There is only one real universe. So if we merely pick a loopy brane at random, the chance of hitting the Goldilocks zone is pretty much zero. You wouldn't bet on a horse at those odds, let alone a universe.

Fortunately, good old quantum gallops to our rescue. Quantum systems can, and do, 'tunnel' from one energy valley to another. The uncertainty principle lets them borrow enough energy to do that, and then pay it back so quickly that the corresponding uncertainty about timing prevents anyone noticing. So, if you wait long enough – umptyplexplexplex years, perhaps, or umptyplexplexplexplex if that's too short – then a single quantum universe will explore every valley in the entire branescape. Along the way, at some stage it finds itself in a Goldilocks valley. Life like ours then arises, and wonders why it's there.

It's not aware of the umptyplexplexplexplex years that have already passed in the multiverse: just of the few billion that have passed since the wandering universe tunnelled its way into the Goldilocks range. Now, and only now, do its human-like inhabitants start to ask why it's possible for them to exist, given such ridiculous odds to the contrary. Eventually, if they're bright enough, they work out that thanks to the branescape and quantum, the true odds are a dead certainty.

It's a beautiful story, even if it turns out to be wrong.

FIFTEEN

AUDITORS
OF REALITY

IT WAS ONE HOUR LATER. Wizards were ranged in rows across the width of the Great Hall in a variety of costumes, but mostly in what might be called Early Trouser; despite Rincewind's view on nudity, a grubby shirt and pants would pass without comment in many ages and countries and lead to fewer arrests.

'Right, then,' said Ridcully, striding along the ranks 'We've kept all this *very* simple so that even *professors* can understand! Ponder Stibbons has given all of you your tasks!' He stopped in front of a middle-aged wizard. 'You, sir, who are you?'

'Don't you know, sir?' said the wizard, taken aback.

'Slipped m' mind, man!' said Ridcully. 'Big university, can't be expected to recognise *everyone*!'

'It's Pennysmart, sir. Professor of Extreme Horticulture.'

'Any good at it?'

'Yes, sir!'

'Any students?'

'No, sir!' said Pennysmart, looking offended.

'That's what I like to hear! And what will you be doin' today?'

'First, it appears, I shall be dropped waist-deep in a lagoon in the, the –' he stopped, and fumbled a piece of paper out of his pocket '– Keeling Islands, where I shall attack the sand bottom round me with this rake,' he held up the implement, 'and then return here as soon as I see any humans.'

'And how will you do that?'

'Say aloud, "Return Me, Hex",' said Pennysmart, smartly.

'Well done, good man,' said the Archchancellor. He raised his voice. 'Remember that, everyone! Exactly those words! Write them down if you can't remember them. Hex will bring you back on the lawn *outside* the building. There will be hundreds of you and many of you have several tasks, so we don't want any collisions! Now, if—'

'Excuse me,' said Pennysmart, raising a hand.

'Yes?'

'*Why* will I be standing in a lagoon flailing around with a rake, please?'

'Because if you don't do that, Darwin will tread on the dorsal spine of an extremely poisonous fish,' said Ponder Stibbons. 'Now—'

'Excuse me again, please,' Pennysmart said.

'Yes?'

'Why won't *I* tread on this fish?'

'Because *you* will be lookin' where you are treadin', Mr Pennysmart,' roared Ridcully.

But a forest of other hands had gone up. About the only wizard without a hand aloft was Rincewind, who was staring gloomily at his feet.

'What's all this about?' said the Archchancellor, irritably.

'Why have I got to move a chair six inches?'

'Why have I got to fill up a hole in the middle of a prairie?'

'Why have I got to hide a pair of trousers?'

'Why have I got to stuff a letter box full of starved snails?'

Ponder waved his clipboard wildly to silence the clamour.

'Because otherwise Darwin would have fallen off a chair or been thrown from a horse or would have been struck by a stone hurled by a rioter or an unwise letter would have reached its destination,' he said. 'But there are more than two thousand tasks, so I can't explain *every* one. Some of them are the start of a quite astonishing causal chain.'

'We are supposed to develop questioning minds, you know,' someone muttered.

'Yes, but not regarding university policy!' said Ridcully. 'You all have very simple jobs to do! Gentlemen, Mr Stibbons will call out your names, and you will step smartly into the circle! Over to you, Mr Stibbons!'

Ponder Stibbons picked up a different clipboard. He was beginning to collect clipboards. They proclaimed order in an increasingly hard-to-understand world. That's all I've ever really wanted, he thought. I just want to feel that things are being ticked off properly.

'Now, chaps,' he said. 'This should not be hard, as the Archchancellor has said. If at all possible don't talk to anyone and don't touch anything. In and out, that's the ticket. I want this done fast. I have a . . . theory about that. So don't waste time, wherever you go. Are we all ready? Very well . . . Aardvarker, Professor A . . . '

One by one, with confidence or trepidation or a mixture of both, wizards stepped into Hex's circle of light and vanished. As they did so, little pointy-hatted wizard symbols appeared at points in the tangle of lights above.

Rincewind watched gloomily, and didn't join in the ragged cheer as, one by one, red circles began to wink out.

Ponder had taken him aside earlier and had explained that, since Rincewind was so experienced at this sort of thing, he was going to be given four of the most, er, interesting tasks. That was how he had put it: 'er, interesting'. Rincewind knew all about 'er, interesting'. There was a giant squid out there with his name on it, that's what it meant.

A movement at the end on the hall made him look around. It was a chest, a metal-bound box of the kind favoured by people who bury treasure, and it walked on hundreds of little pink legs. He groaned. He'd left it asleep on the wardrobe in his bedroom, with its feet in the air.

'Hmm?' he said.

'Rincewind! Off you go, best of luck!' Ponder repeated. 'Hurry up!'

There was nothing for it. Rincewind walked into the circle, and fell over as the ship moved gently under him.

It was dawn, and a clammy sea mist was drifting across the deck. Rigging creaked, the water lapped far below. There was no other sound. The air smelled warm and exotic.

There was a small cannon only a few feet away. Rincewind knew about them. He was the only wizard to have seen one, over in the Agatean Empire, where they were known as 'Barking Dogs'. He was sure that one of the rules associated with them was 'do not stand in front'.

Slowly, he reached inside his shirt and pulled out his pointy hat. It was red, or rather, it was the colour that red becomes after being washed, eaten, retrieved, scorched, buried, crushed, engulfed, washed again and wrung out far, far too often.

No wearing of pointy hats? Were they mad? He pulled at it a bit to get it back to its comfortable shapeless shape, and put it on. That felt much better. A pointy hat meant you weren't just *anyone*.

He unrolled his instructions.

1. Remove ball from 'cannon'

There was no one around. There *was* a stack of metal balls by the cannon. Rincewind pulled the barrel around with some effort, felt down the hole, and grunted as his fingers touched the top of another ball at the far end.

How could he get it out? The way to get a ball out of a Barking Dog was to set a match to its tail, but Ponder had said this wasn't an option. He cast around, and saw a bundle of tools by the stack; one was a rod with an end like a super-corkscrew.

Carefully, he pushed it down the cannon, wincing at every *clink*. Twice he felt the curved springy bits engage with the ball, and twice it came away and rolled back with a thud.

At the third attempt he was able to get the tapped ball almost out of the mouth of the barrel, and slid his fingers under it.

Well, that wasn't too hard, was it? He dropped it over the side, where the sea swallowed it with a 'plomp!'

This caused no stir anywhere. Job done, and nothing horrible had happened at all! He pulled a scrap of paper out of his pocket. It was important to get the words right.

'Return—' he began, and stopped. With a little metallic grinding noise, another ball rolled gently off the pile, across the deck, and leapt into the cannon's mouth.

'O-kay,' said Rincewind slowly. Of course. Obviously. Why had he thought otherwise for even one second?

Sighing, he picked up the ball grasper, rammed it down the barrel, caught the ball, and jerked it out so hard that it would have made a giveaway noise hitting the deck. Fortunately, it landed on Rincewind's foot.

A little metallic sound disturbed him while he was lying across the barrel making the traditional 'gheeee' noise of those who are screaming through clenched teeth.

It was the noise of another cannon ball rolling across the deck. He jumped on it, picked it up, and felt a slight resistance trying to tug it out of his hands. He wrenched against the invisible force, spun around and the ball flew out of his hands and over the rail.

This time the 'plomp!' caused an interrogatory mumble from below decks.

The last remaining ball started to roll towards the cannon.

'Oh no you don't!' snarled Rincewind, and grabbed it. Again the force tried to pull the ball away from him, but he clung on tightly.

There was the sound of footsteps climbing stairs. Somewhere close, in the fog, someone sounded angry.

Then in the billows in front of Rincewind there was . . . something. He couldn't make out the shape, but it disturbed the fog, making an outline of sorts. It looked like—

It let go as someone hurried closer. Rincewind growled in triumph, staggered backwards, tipped over the rail and, still clutching the cannon ball, went 'plomp!'

'Look at the red circles, sir!' shouted Ponder.

Erratically, in the drifting tangle of lights, the red circles were winking out. The yellow line was extending.

'That's the style, Mr Stibbons!' the Archchancellor roared. 'Keep pounding away!'

Wizards were scuttling through the hall, getting fresh instructions, catching their breath and disappearing in the circle again.

Ridcully nodded at the stretcher containing the screaming Pennysmart, as it was hurried away to the Infirmary.

'Never seen *that* shade of purple on a leg,' he said. 'I *told* him to look where he was going. You heard me, didn't you?'

'He says he was dropped right on top of the fish,' said Ponder. 'I'm afraid Hex is running at the very limit of his power, sir. We're bending an entire timeline. You've got to expect some accidents. A few of the returning wizards are reappearing in the fountain. We just have to accept that it's better than them reappearing inside walls.'

Ridcully surveyed the throng, and said: 'Here comes one from the fountain, by the look of it . . . '

Rincewind limped in, his face like thunder, water still streaming off him, with something grasped in his hands. Halfway across the hall a fish fell out of his robe, in obedience to the unbreakable laws of humour.

He reached Ponder, and dropped the cannon ball on the floor.

'Do you know how hard it is to shout underwater?' he demanded.

'But I see you were successful, Rincewind,' said Ridcully.

Rincewind looked up. All over the streaming lines, little pointy wizard symbols were appearing and disappearing.

'No one told me it would fight back! It fought back! The cannon tried to load itself!'

'Aha!' said Ridcully. 'The enemy is revealed! We're nearly there! If they are breaking the—'

'It was an Auditor,' said Rincewind, flatly. 'It was trying to be invisible but I saw it outlined in the fog.'

Ridcully sagged a little. A certain exuberance faded from his face. He said, 'Oh, darn,' because an amusing misunderstanding in his youth had led him to believe that this was the worst possible word you could say.

'We've found no evidence of them,' said Ponder Stibbons.

'Here? Did we look? We wouldn't find any anyway, would we?' said Ridcully. 'They'd show up as natural forces.'

'But how could they exist here? All those things work by themselves here!'

'Same way we did?' said Rincewind. 'And they'll meddle with anything. You know them. And they really, really hate people . . .'

Auditors: personifications of things that have no personality that can be imagined. Wind and rain are animate, and thus have gods. But the personification of gravity, for example, is an Auditor or, rather Auditors. In universes that run on narrativium rather than automatic, they are the means by which the most basic things happen.

Auditors are not only unimaginative, they find it impossible to imagine what imagination is.

They are never found in groups of less than three, at least for long. In ones and twos they quickly develop personality traits that make them *different*, which to them is fatal. For an Auditor to have an opinion that differs from that of its colleagues is certain . . . cessation. But while individual Auditors cannot hold an opinion (because that would make them individual), Auditors as a whole certainly can, and with grim certainty they hold that the multiverse would be a lot better off with no life in it. Life gets in the way, tends to be messy, acts unpredictably and reverses entropy.

Life, they believe, is an unwanted by-product. The multiverse would be more reliable if there wasn't any. Unfortunately, there are rules. Gravity is not allowed to increase a millionfold and laminate all local life forms to the bedrock, highly desirable though that would appear to be. Simply mugging life forms merely walking, flying, swimming or oozing past would attract attention from higher authority, which Auditors dread.

They are weak, not very clever and always afraid. But they can be subtle. And the wonderful thing about intelligent life, they have discovered, is that with some care it can be persuaded to destroy *itself*.

SIXTEEN
MANIFEST
DESTINY

 THE WIZARDS ARE DISCOVERING THAT changing history is not so easy, even when you've got a time machine. The Auditors aren't helping, but history has its own metaphorical Auditor, often called 'historical inertia'. Inertia is the innate tendency of moving objects to continue moving along much the same track, even if you try to divert them; it is a consequence of Newton's laws of motion. Historical inertia has a similar effect but a different cause: changing a single historical event, however important it may appear, may have no significant effect on the social context that directs the path of history.

Imagine we've *got* a time machine, and go back to the past. Not too far, just to the assassination of Abraham Lincoln. In *our* history, the President lived till the following morning, so a tiny deflection of the assassin's bullet could make all the difference. So we arrange a small deflection, and he is hit but recovers, with no noticeable brain damage. He cuts a couple of appointments while he recuperates, and then he goes on to do ... what?

We don't know *anything* about that new version of history.

Or do we? Of course we do. He doesn't turn into a hippopotamus, for a start, or a Ford Model T. Or disappear. He goes on being President Abraham Lincoln, hedged in by all the political expediencies and impossibilities that existed in our version of history *and still exist in his.*

The counterfactual* scenario of a live Lincoln raises many questions. How much do you think being the American President is like driving a car, going where you want to? Or sitting in a train, observing the terrain that others drive you through?

Somewhere in between, no doubt.

Ordinarily, we don't have to think much about counterfactuals, precisely because they are contrary to fact. But mathematicians think about them all the time – 'if what I think happens is wrong, what can I deduce that might *prove* it wrong?' Any consideration of phase spaces automatically gets tangled up in worlds of if. You don't really understand history unless you can take a stab at what might have happened if some major historical event had not occurred. That's a good way to appreciate the significance of that event, for a start.

In that spirit, let's think about that altered 'now': the beginning of the West's third millennium of history, but without Lincoln having been assassinated in its past. What would your morning newspaper be called? Would it be different? Would you still be having much the same breakfast ritual, bacon and eggs and a sausage perhaps? What about the World Wars? Hiroshima?

A very large number of stories have been written with this kind of theme: Wilson Tucker's *The Lincoln Hunters* is set in such an 'alternat(iv)e universe' and tackles the Lincoln question.

Curious things happen in our minds when they are presented with *any* fictional world. Consider for a moment the London of the late nineteenth century. It did have Jack the Ripper, and we can wonder about the real-world puzzle of who he was. It had Darwin, Huxley and Wallace, too. But it did *not* have Sherlock Holmes, Dracula, Nicholas Nickleby, or Mr Polly. Nevertheless, some of the best

---

* Counterfactual: a more acceptable word for what has for a long time been a very common feature of science fiction, the 'alternate world' or 'worlds of if' story (there was a pulp SF magazine in the 1950s called *Worlds of If*, in fact). 'Counterfactual' is now used when said stories are written by real writers and historians, to save them the indignity of sharing a genre with all those strange sci-fi people.

portrayals of the Victorian world are centred around those characters. Sometimes the fictional portrayals are intended to paint a humorous gloss on the society of the period. The Flintstones put just such a gloss on human prehistory, so much so that in order to think rationally about our evolution we must excise all those images, which is probably an impossible task.

Sherlock Holmes and Mr Polly were Victorians in just the same sense that the tyrannosaur and triceratops in *Jurassic Park* were dinosaurs. When we envisage *Triceratops*, we cannot avoid the memory of that warty purple-spotted *Jurassic Park* skin, as the beast lies on its side, breathing stertorously. And *Tyrannosaur*, in our mind's eye, is running after the jeep, bobbing its head like a bird. When we envisage late nineteenth-century Baker Street it's very difficult not to see Holmes and Watson (probably in one of their filmic versions) hailing a four-wheeler, off to solve another crime. Our pictures of the past are a mixture of real historical figures and scenarios peopled by fictional entities, and it's difficult to keep them apart, especially as films and TV series acquire better technologies to latch into those spurious pictures in our heads.

The 1930s philosopher George Herbert Mead made much of the rather obvious point that the present, in a causal world, does not only determine ('constrain' if you prefer) the future, it also affects the past, in just this sense: if I discover a new fact about the present, then the (conceptual) past that led up to the new present must also have been different. Mead thereby enabled a rather cute way of seeing how good the portrayals of Sherlock Holmes, or of the *Jurassic Park* tyrannosaur, are. If my picture of the present isn't altered at all by the presence or absence of Sherlock Holmes in the 1880s, or if my construction of the present by evolutionary processes isn't altered at all by seeing *Jurassic Park*, then these are *consistent* inventions.

Dracula and the Flintstones are *inconsistent* inventions: if they really existed in our past, then the present isn't what we think it is. Much of the fun of 'worlds of if' stories, and of many consistent fictions like *The Three Musketeers*, is that they show closed-loop causalities in our apparent past. Whether or not D'Artagnan had aggregated the Musketeers and thereby brought into being much of the causal history of seventeenth-century France, children of later centuries would learn the same history in the textbooks. Ultimately, consistent historical fictions make no difference.

In *The Science of Discworld II* we played with this idea in several ways: the presence of the Elves was, surprisingly, consistent with our history; stopping them led to stagnation of humans and had to be reversed. In this book the meddling of the Unseen University wizards, in Victorian history this time, is trying to create an apparently internally caused history in which Darwin wrote *The Origin of Species* and not *Theology of Species*. We are going to use this trick to illuminate the causalities of human history.

In order to do this convincingly, we must make the Discworld intrusions consistent, but even then we must address the convergence/ divergence problem, which is this. Would such a meddled-with world *converge* on to ours, demonstrating that history is stable, or would any tiny difference start a *divergence* that became wider and wider, proving history to be unstable?

Most people think the latter. Indeed, even the wildly imaginative physicists who believe that a new world history is created by each and every decision in this universe, spawning new universes in which the *other* choices were implemented, don't imagine that the histories converge. No, each universe goes its own way, spitting out new and divergent universes as it goes. The Trousers of Time are a tree: their legs can branch but never merge.

The *Worlds of If* stories were divided on this issue. Some had each tiny change in the past getting amplified, resulting in vast changes now: we've mentioned Bradbury's story where you trod on a butterfly in the far past, on a dinosaur hunt, and came back to find a fascist regime. Or the changes you made were all wiped out, because there was a gigantic all-powerful inertia-of-events Kismet that you couldn't change. However you tried to avoid your fate, that only made it more certain to happen. And some stories took a middle way; some things converged and others didn't.

This, we think, is the rational way to think about time travel and altering the past.

After all, we don't change the *rules* by which the past works. Gravity still operates, sodium chloride crystals are still cubical, people fall in and out of love, misers hoard and spendthrifts squander. What we change is what physicists call the 'initial conditions'. We change the positions of a few of the pieces on the Great Chessboard of Life, The Universe and Everything, but we still keep to the rules of chess. That's how the wizards operated in *The Science of Discworld II*. They went back in time to remove the Elves from the game board; then they went back *again* to stop themselves making that mistake.

We are now ready to think about our question above: would the names of newspapers have changed if Abraham Lincoln had lived to a ripe old age?

Perhaps some of them would, because some cultures would have become rather different. Perhaps Quebec wouldn't have been French; perhaps New York would have been Dutch. But names like *Daily Mail*, *Daily News* and *New York Times* are so obvious, so appropriate, that even if the Roman Empire were still running things, the Latin equivalents would seem fitting. Someone would have invented flush toilets, and there would have been a steam engine time, when several people invented steam power. Some things in Western culture seem so *likely*, from toilet paper on up to (as soon as paper is invented) daily newspapers to plastics to artificial wood . . . Technology seems to have a set of rules for its advancement, so that it seems rational to expect gramophones of some kind if people make music with musical instruments, then tape players when people get used to electricity and its possibilities for amplification. Then from analogue to digital, to computers . . . some things seem inevitable.

Perhaps this feeling is misleading, but it's silly to insist that absolutely *everything* in a slightly divergent future has to end up different.

Organic evolution has lessons for us here, and these lessons can instruct us about how likely various advances in animal organisation were. Innovations like insect wings, vertebrate jaws, photosynthesis, life coming out from the seas on to the land . . . if we ran evolution on Earth again, would the same things happen? If we went back to the beginning of life on this planet, and killed it, would another system evolve and give us a whole different range of creatures, or would Earth remain lifeless? Or would we be unable to decide whether we'd done anything, because everything would be just the same the second time around?

If history 'healed up', we wouldn't be able to tell if it was the second, or the hundredth, or the millionth time around – each time sooner or later producing a version of us, whose time machine goes back to *The Origin*. There would be a consistent time loop, as happened with the Elves in *The Science of Discworld II*. If life is 'easy' to originate (and the evidence does look that way) then this isn't an exercise in going back and killing your grandfather, or if it is, your grandfather is a vampire and doesn't *remain* killed. If life is easy to invent, then preventing it happening once, or a million times, will make no difference in the long run. The same process that generated it will happen again.

Looking at the panorama of life on this planet, in time as well as space, we can see that there are two kinds of evolutionary innovation. Photosynthesis, flight, fur, sex, and jointed limbs have all arisen independently in several different lineages. Surely, like toilet paper, we would expect to see them again each time we ran life on Earth.

And, presumably, we'll see them on other aqueous planets when we explore our local region of the galaxy. Such evolutionary attractors are called 'universals', in contrast to 'parochials': unlikely innovations that have happened only once in Earth's history.

The classic parochial is the curious suite of characters possessed by land vertebrates, because a particular species of Devonian fish succeeded in invading the land in our, *real*, history. Those fishes' descendants were amphibians, reptiles, birds, and mammals – including us. Jointed limbs are a universal innovation. The limbs of spiders, hydraulically operated, differ in detail from the limbs of mammals, and were presumably acquired via a different ancestor, perhaps an earlier arthropod proto-spider. The mammalian internal skeleton, with one bone at the body end, then two, then a wrist or ankle, then five lines of bones for fingers or toes, was an independent evolution of the same universal trick.

This highly unlikely combination now occurs in all land vertebrates (except most of the legless ones), because they are all descended from those fishes that came out of the water to colonise the land. Other parochials are feathers and teeth (of the kind that evolved from scales, which are what we have). And, especially, each of the special body-plans that characterise Earth's animals and plants: mammal, insect, rotifer, trilobite, squid, conifer, orchid . . . *None* of these would appear again after a rerun of Earth's evolutionary history, nor would we find exact replicas on other aqueous planets.

We would expect much the same *processes* to occur, though, in a repeat run of Earth or on another similar world: an atmosphere far from chemical equilibrium as life forms pump up their chemistry using light; planktonic layers of the seas colonised by the larvae of sedentary animals; flying creatures of many kinds. Such ecosystems would also probably have 'layers', a hierarchical structure, fundamentally similar to the ecosystems that have emerged in so many different circumstances on Earth. So there would be 'plantlike' creatures, a productive majority of the biomass (like Earth's grass or marine algae). These would be browsed by tiny animals (mites, grasshoppers) and by larger animals (rabbits, antelopes), with a few very large creatures (elephants, whales). Comparable evolutionary histories would lead to the same dramatic scenarios, but performed by different actors.

The central lesson is that although natural selection has a very varied base to work with (recombinations of ancient mutations, differently assorted in all those 'waste' progeny), clear large-scale themes emerge. Marine predators, such as sharks, dolphins, and ichthyosaurs all have much the same shape as barracuda, because hydrodynamic efficiency dictates that streamlining will catch you more prey, more cheaply. Very different lineages of planktonic larvae all have long spines or other extensions of the body to restrain the tendency to fall or rise because their density differs from that of seawater, and most of them pump ions in or out to adjust their densities too. As soon as creatures acquire blood systems, other creatures – leeches, fleas, mosquitoes – develop puncture tools to exploit them, and tiny parasites exploit both the blood as food and the bloodsuckers as postal systems. Examples are malaria, sleeping-sickness, and leishmaniasis in humans, and lots of other parasitic diseases in reptiles, fishes, and octopuses.

Large-scale themes may be the obvious lesson, but the last examples reveal a more important one: organisms mostly form their own environments, and nearly all of the important context for organisms is other organisms.

Human social history is like evolutionary history. We like to organise it into stories, but that's not how it really works. History, too, can be convergent or divergent. It seems quite sensible to believe that small changes mostly get smeared out, or lost in the noise, so that big changes are needed to divert the course of history. But anyone familiar with chaos theory will also expect some tiny differences to set off divergent histories, drifting progressively further away from what might have happened otherwise.

Changing history is a theme of time-travel stories, and the two issues come together in those stories called 'worlds of if'.

We have the strongest feeling that what we do, even what we decide, does change history. If I decide, now, not to go and meet Auntie Janie at the train station even though she's expecting me because I told her I would . . . the universe will take a different path from the one it would have taken if I had done the expected. But we've just seen that even saving Abraham Lincoln from the assassin would have the tiniest, most local, of effects. Neighbours such as the gas-bag aliens on Jupiter wouldn't notice Lincoln's survival at all, or at least not for a very long time. After all, we haven't yet noticed *them.**

In fact, how *will* they, or we, notice? How will we be able to say, 'Just a minute, this newspaper shouldn't be called the *Daily Echo* . . . There must have been a time traveller interfering, so that we're now in the wrong leg of the Trousers of Time'?

Auntie Janie making her own way from the station won't topple empires – unless you believe, with Francis Thompson's *The Mistress of Vision*, that

> *All things by immortal power*
> *Near or far*
> *Hiddenly*
> *To each other linkéd are*
> *That thou canst not stir a flower*
> *Without troubling of a star.*

That is, all contingent chaos butterflies are responsible in some sense for all important events like hurricanes and typhoons – and

* Well, there *might* be . . .

newspaper titles. When a typhoon, or a newspaper tycoon, topples an empire, that event is caused by everything, all those butterflies, that preceded it. Because change in any one – or perhaps just in one of a very large number – can derail the important event.

So everything must be caused by *everything* before it, not just by a thin string of causality.

We think about causality as a thin string, a linear chain of events, link following link following link . . . probably because that's the only way we can hold any kind of causal sequence in our minds. As we'll see, that's how we deal with our own memories and intentions, but none of this means that the universe can isolate such a causal string antecedent to any event at all, important or not. And surely 'important' or 'trivial' is usually human judgement, unless the universe really does 'smear out' most small changes (whatever that means), and major events are those whose singular influence can be distinguished at later times.

Because they *are* stories, committed to the way our minds work and not to the way the universe works its own causality, most time-travel stories assume that a big (localised) change is needed to have a big effect – kill Napoleon, invade China . . . or save Lincoln. And time travel stories have another convention, another 'conceit', because they are stories, nearer fee-fi-fo-fum than physics. This is the remembered timeline of the traveller. Usually the plot depends on it being unique to him. When he comes back to his present *he* remembers stepping on the butterfly, or killing his grandfather, or telling Leonardo about submarines . . . but no one else is conscious of anything other than their 'altered' present.

Let's move from large events, large or small causes, to how we influence the apparent causality in our own lives. We have invented a very strange oxymoron to describe this: 'free will'. These words appear prominently on the label of the can of worms called 'determinism'. In *Figments of Reality* we titled the free will chapter: 'We wanted to have a chapter on free will, but we decided not to, so here it is' in order to expose the paradoxical nature of the whole idea. Dennett's recent book *Freedom Evolves* is a very powerful treatment of the same topic. He shows that in regard to 'free will' it doesn't *matter* whether the universe, including humans, is deterministic. Even if we can do only what we must, there are ways to make the inevitable evitable. Even if it is all butterflies, if tiny differences chaotically determine large historical trends, nevertheless creatures as evolved as us can have 'the only free will worth having', according to Dennett. He writes of dodging a baseball coming for his face, and this being perhaps a culmination of a causal chain going right back to the Big Bang – yet if it will help his team, he *might* let it hit his face.

But then, what decides it is: will it help his team? That's not a free choice.

Inevitable, evitable.

Dennett's best example is more ancient: Odysseus's ship approaching the Sirens. Inevitably, if his men hear the Sirens' song, they will steer the ship on to the rocks. But the steersman must be able to hear the surf, so there seems no way to avoid their lure. Odysseus has himself lashed to the mast, while all his sailors plug their ears with wax so they cannot hear the Sirens. The vital issue for Dennett is that humans, and on this planet probably only humans, have evolved several stages beyond the observing-and-reacting that even quite advanced animals do. We observed ourselves and others observing, so got more context to embed our behaviour in – including our prospective behaviour. Then we developed a tactic of labelling good and bad imaginary outcomes, just as we labelled our memories with emotional tags. We, and some other apes – perhaps also dolphins, perhaps even some parrots – developed a 'theory of mind', a way to imagine ourselves or others in invented scenarios and to anticipate the associated feelings and responses. Then we learned to run more than one scenario: 'But on the other hand, if we did so-and-so, the lion couldn't get us anyway...', and that trick soon became a major part of our survival strategy. So with Odysseus . . . and fiction . . . and particularly that dissection of hypothetical alternatives that we call a time-travel story.

In our minds, we can hold many possible histories, just as Mead showed that every discovery about today implies a different past leading up to it. But whether there is any sense in which the universe has several possible pasts (or futures) is a much more difficult question. We've argued that popularisations of quantum indeterminacy, particularly the many-worlds model, have got confused about this. They tell us that the *universe* branches at every decision point, whereas we think that people have to invent a different *mental* causal path, a different explanatory history, for each possible present or future.

Antonio Damasio has written three books: *Looking for Spinoza*, *Descartes' Error*, and *The Feeling of What Happens*. These are popular accounts of what we know about the important attributes of our minds. He has documented our discoveries, now that we can use various experimental techniques to 'watch the brain thinking' and see how the different parts of the brain are involved in what we feel about the things we think. We tend to forget that our brains are continually interacting with our bodies, which supply the brain with stance-determining hormones for longer-term behaviour, and mood-changing emotion-provoking chemicals for short-term modulation of our intentions and feelings, directing our thoughts.

According to these books, the result of having lived with a brain which we think we direct using a kind of tiller, but which actually is continually affected by cross-winds, occasional storms, rain and warm sun that provokes us into lazy days, is that we have evolved a series of memories with different flavours. Or, the result of having lived with a brain that we think we direct using a kind of automobile steering wheel and foot controls, but whose route is actually continually affected by long-term goals that change ('Let's go to a hotel, not to Auntie Janie's *again*'), short-term road signs and other traffic, is that we have evolved a series of memories with different flavours. Or, each of us has a personal history which we explain internally by feelings attached to emotional memories, so we have evolved a series of memories with different flavours.

Damasio has imported emotional biasing into how we think about our own intentions, choices, other people, memories, and prospective plans. He claims that this is what emotion is 'for', and most psychologists now agree that emotionally labelled memories are the effect of having a brain whose interaction with its body paints emotions on to memories and intentions.

We habitually assume that real physical history, and particularly social history, works the same way as our own personal histories, with events labelled 'good' or 'bad' . . . but it doesn't. It's misleading to think of the Big Bang, for example, as an explosion like a bomb or a firework, *seen from outside*. The whole point of the Big Bang metaphor is that at the moment the universe was born, there *was* no outside. More subtly, perhaps, we tend to think of the birth of the universe in the same way that we think of our own birth, or even our conception.

Real history, *post* whatever the Big Bang 'really' was, relies on the accumulation of countless tiny sequences of cause-and-effect. As soon as we begin to think about what any of these sequences looks like, taking it out of the context that drives it, we lose its causality. This seething sea of processes and appearances and disappearances, where no causality can be isolated, is sometimes called 'Ant Country'. The name reflects three features: the seething, apparently purposeless activity of ants, which, in aggregate, makes ant colonies work; the metaphorical Aunt Hillary in Douglas Hofstadter's *Gödel, Escher, Bach*, who was a sentient anthill and recognised the approach of her friend the anteater because some of her constituent ants panicked; and Langton's Ant, a simple cellular automaton, which shows that even if we know all the rules that govern a system, its behaviour cannot be predicted except by running the rules and seeing what happens. Which in most people's book is not 'prediction' at all.

For similar reasons, it is impossible to forecast the weather accurately, even a few weeks ahead. Yet, despite this apparent absence of causality at the micro-levels of weather, the impossibility of isolating causality in the swirling butterflies ... despite the chaotic nature of meteorology in both the large and the small, weather *makes sense*. So does a stone tumbling downhill. So does a lot of physics, engineering, and aeronautics: we can build a Boeing 747 that flies reliably. Nevertheless, all of our physical models are rooted in brains that get most of their perceptions wrong.

Shouting at the monkeys in the next tree. That's what brains evolved to do. Not mathematics and physics.

We get ecology and evolution mostly right, but often wrong, for the same reasons. The scenarios we build don't work, they're as false to fact as 'weather'. But we can't help building them, and they're useful sufficiently often to be 'good enough for government work'.

To underline this point, here's an important evolutionary example. Think of the first land vertebrate, that fish that came out of the water. We have the strongest feeling that if we took a time machine back to the Devonian, when that first important fish was emerging from the sea, there ought to be a moment that we could isolate: 'Look, by wriggling out on to the mud *that* female has escaped that predator, so she's lived to lay her eggs, and some of them will become our ancestors ... If she hadn't got those leggy fins, she wouldn't quite have made it, and we wouldn't be here.'

Grandfather paradox again? Not quite, but we can illuminate the grandfather paradox neatly with this example. Ask yourself what would happen if you killed *that* fish. Would humanity never have happened? Not at all. By isolating a single event, we have tried mentally to make history follow a thin thread of causality. But we made the Adam-and-Eve mistake: ancestors don't get *fewer* as you go back, they multiply. You have two parents, four grandparents, maybe only seven great-grand parents, because cousin marriages were commoner then. By the time you've gone back a couple of dozen generations, a significant proportion of all the breeders of that period were your ancestors. That's why everyone finds some famous ancestors when they look – and the fact that famous people were rich and powerful and sexually successful helps too, so that they are reproductively better represented in that generation's descendants.

Note that we said 'breeders' and 'many'. Nearly all sexually produced creatures don't breed, including humans of most previous generations. Not only are most of the people alive at that previous generation young children who won't survive to breed; many of the apparently successful breeders contribute to lineages that die out before they get to the present day, because they are excluded from the limited ecosystem by more successful lineages as the generations pass.

So when we look at those Devonian fishes, there simply isn't just *one* that was our ancestor. *All* of the breeders, a very unsystematic small proportion of the fish population, contributed to the recombining and mutating mix of genes that passed down from those fishes that left the water, through generations of amphibians and mammal-like reptiles, into the early mammals, were newly selected to characterise the early primates, and eventually ended up in us. There wasn't a single grandfather fish, or one grandfather primate, no thin line of descent, just as there isn't a thin line of causality leading from a butterfly's wing flap to a hurricane. Nearly any fish you went back and killed would make virtually no difference to history. We'd still be here, but history would have taken a slightly different route to get to us.

But that doesn't mean that history has no important accomplishments.

Some physicists, especially, have argued from this indeterminacy and chaotic influences at all the micro-levels that there is no pattern to history, that Heisenberg uncertainty rules. Wrong. Just because we cannot predict the weather more than about a week ahead, with the best and biggest computers, doesn't mean that there isn't such a thing as weather. Our thin-causal-thread evolutionary scenarios for the emergence of those fishes on to the land don't work, but that doesn't mean we must throw away all ideas of causality in evolution. Any event, when looked at in detail, seems not to have a clear cause, but that just means that our Damasio-minds are not suited to that way of analysing history.

We are much better at totally disregarding all the micro stuff, and making big guesses: I guess it'll be sunny again tomorrow; or I guess that among all those fishes eating each other on the Devonian mudflats, some will escape on to the land. We're confirmed in that guess by finding climbing perch, mudskippers and lots of other separate fish lineages doing exactly that on mudflats today.

The great evolutionary biologist Stephen Jay Gould got this point wrong in *Wonderful Life*: if evolution ran again, he stated, we would not get people, because of all the tiny chaotic butterflies that determined evolutionary outcomes, so there were no thin causal threads. We disagree: we might not, almost surely would not, get the same primate coming down from the trees, but equivalent major innovations would occur in the new and different lineages. People are good at finding high-level groupings, making analogies and metaphors, arguing from what Aunt Janie does today to what she'll do tomorrow, or did twenty years ago. But we oversimplify when we try to disentangle the maze of tiny causalities that lies behind any historical event, because we can't handle that kind of complexity.

So, even though all of the causality happens at the micro-level, and we can't analyse it except in terms of tens of particles interacting when it's really billions, this isn't what it's about. It's like the early twentieth-century physicists telling us that the dining-room table wasn't really there, it was nearly all empty space, and that concepts like 'hard' and 'brown' had no place in the physicist's view of the world. So much the worse for the physicist. Did he really not eat his dinner off just such a hard, brown table? And was not his brain designed to do really clever things with abstractions useful in his daily life, like hard and brown, rather than the very peculiarly unuseful concepts of atoms, nuclei, and so on?

On the contrary, our brains are excellent at all the higher-level judgements they're called on to make, especially in a world that is full of hard, brown tables, doors, houses, trees to make them out of, and other people to help us or compete with us. But nearly all human brains are poor when it comes to the physics of atoms and the micro-world.

Back to history. We 'make sense' of large movements like the Enlightenment, democracy in ancient Athens, the Tudors; but we know that if we were to look at all the small-scale interactions, they would make little sense against the comprehensible backdrop. That is precisely why historical novels can be so fascinating, and why *The Three Musketeers* didn't really affect Cardinal Richelieu and all the important people in seventeenth-century France. Nevertheless, we greatly enjoy the fiction that makes sense of the great movements by tying them down to the motives and nobility of a few people like D'Artagnan, with whom we can identify. The sequels *Ten Years After* and *Twenty Years Later* intrigued some of us, as Dumas found that he was on to a good thing and turned out more of the same. Some of us, at least, then found that Athos's nobility rang increasingly false, and Porthos's good humour was boring, while Aramis's religiosity wore very thin as the years rushed by. The initial idea wedged into the history we knew, it was consistent and provided colourful incident. But the later money-spinners were increasingly at odds with how we knew history worked.

There is an excellent example of the converse of this, which makes that point even better than Dumas. Wells's *The Time Machine*, as we've said, was the absolute classic time travelogue, showing us the large picture from prehistory to the social consequences of the capitalism that the socialist Wells wanted to criticise. Then the cooling Sun, the great crabs on a post-diluvian beach . . . lovely. But Stephen Baxter's modern sequel *The Time Ships* shows us how clever the Morlocks will be, how the Traveller is really a little bit prurient about the little girl from the future – a resonance with Lewis Carrol's Alice – who is innocent and a bit stupid.

It's like a historical novel that puts all the little sexy and despicable bits into the great tapestry of history. Such literary exercises add colour and flavour to history, just as Damasio has shown that we do with our own personal memories. The pleasure this exercise gives us shows how our human minds read history: in the large without flavour, in the small with the kind of colour that we paint on to our own small reminiscences. So historical romance is just that: romantic painting of the little, interesting items, whose causality might affect the big picture, but doesn't.

What does it mean, then, to ask whether time knits up any changes, or whether mischievous butterflies are ultimately responsible for the fall of empires?

Here fictional conventions cease to fit the real world. From the point of view of the wizards, Roundworld time is a one-dimensional sequence that they can access two-dimensionally like a book. For narrative reasons, we have to depict it like this because of all those thin-thread-of-causality historical stories that our minds find so congenial. In a fictional context, we have little choice. However, here we want to think about the nature of causality and free will in the 'real' universe, which – as we've made clear throughout the *Science of Discworld* series – does not have any narrativium. In that context, we have to understand that this simple image of Roundworld history is a fraud. The Trousers of Time also work well as a story, but as genuine physics they are a fraud: you can't be pushed from one leg to the other by an event. Worse, you can't tell that there has been such an event. As far as you are concerned, this is the world. It doesn't have 'ifs' in its past.

None of this stops us using 'what ifs' (which by nature are fictions, not facts) to think about history. We can still ask, in our minds, what would have happened if, say, Lincoln had survived . . . but in the real world he didn't, and we can't run a mock-up of 'if he did' in the real world: only in our heads.

Science runs into precisely this difficulty. For instance, the main problem in testing medical treatments is that we can't both *give* Mrs Jones the treatment and *not* give Mrs Jones the treatment, simultaneously, and compare the results. We can do it sequentially, but then the second treatment (whether it is placebo or real treatment) is of a different Mrs Jones, one who's had the first treatment. So what the testers do is to have quite a large panel, do treatment first on some, placebo first on others – and they should perhaps do two placebos on a few, and two treatments on a few others.

What time-travel stories do, in our minds, is the same kind of test: 'What would happen if Leonardo had really seen a submarine working?' or equivalently 'Did Leonardo see a submarine working?' In *The Science of Discworld*, and more explicitly in *The Science of Discworld II*, we asked whether the interesting stories that we make up have some kind of coherent explanation, something like 'evil' – which we personified in the second book as Elves. To what extent do such concepts relate to the real rules of the real world? Now we argue that we *cannot know* if any answer we get is useful; we can't even know whether we've got an answer at all. And that this is precisely why Dennett's kind of free will is the only one worth having. It's prospective, giving each of us the chance to make little items of an otherwise inevitable future evitable.

When we look back on something we've changed by that kind of an act of free will, it's just as causal as everything else – and if the universe is in any sense determinate, then it is determinate in that sense. Think of Odysseus looking back at what happened as his ship failed to be caught by the Sirens. His men didn't hear them, and he, who could hear them, couldn't act to steer the ship. So he and his crew came through in that most unlikely of passages. There is a sense, of course, in which every sea passage is equally unique, just as every deal of cards is unique; but Odysseus's journey, like a one-suit-per-player deal of cards, is totally remarkable too. Looking back into history, can we find journeys, events, and processes so remarkable that they seem to be the results of previous acts of free will?

What, then, is causality? For Damasio-like reasons, we tend to think that what gives history its dynamic is the big events, the 'pivot points'. The fallacy is that we think big causes are needed to produce big effects. This is false (butterfly) but there is a problem: choosing the right tiny change (which butterfly?). And there are always billions of new butterflies, dragging new changes out from previously invisible differences 'in the 13th decimal place', unobservable until their effects show up.

Real history is like this; causes are often distributed, with huge numbers of tiny events all coming together. It is just this problem that leads Ridcully to employ such a huge number of wizards, doing such a bizarre set of trivial things, merely to get *The Origin* written.

We only justify this sort of causality in retrospect: history didn't know 'where it was going'. So changing the past creates a *context* for the future, not a causal chain, and this is how the wizards must operate, which is why we have thousands of them making endless trivial changes to Victorian history, instead of, say, assassinating Queen Victoria. Any Victorian, perhaps particularly the well-trained nursemaid, will tell you just that about your personal history: your heart must be pure (context) rather than your plans being subtle.

SEVENTEEN

GALÁPAGOS ENCOUNTER

 CHARLES DARWIN WAS SITTING ON a grassy bank. Three types of bee buzzed among the flowers, and overhead examples of *Hirundo rustica* swooped after miscellaneous *Ephemeroptera*.

His thoughts were complex, as human thoughts tend to be when the mind is idling, but included: this is an interesting bank of astonishing complexity; there might be fish for lunch; he had a sore throat; he hoped never to receive another letter about barnacles; the rash seemed to be getting worse; there was a strange buzzing sound; had he really experienced that apparition?; homeopathy transcended all common sense; he really should find out where the ovaria were situated in *Phyllosoma;* it really was a *very* loud buzzing . . .

Something like a yellow-brown smoke was issuing from a hole in the bank a few yards away, and resolved itself into a cloud of angry *Vespula vulgaris*. It bore down on the horrified Darwin –

'Over here, waspies!'

Darwin stared.

This mission had created a difficult decision for Rincewind, when he'd been presented with the task of preventing Charles Darwin being stung to death by wasps. Right from the start it was obvious that Darwin would see him, and if Rincewind was invisible the wasps wouldn't see *him*. He'd therefore undertaken the mission carrying two buckets of warm jam and wearing a pink tutu, an acid-green wig and a red nose, reasoning that (a) Darwin wouldn't believe that he had seen him and in any case (b) wouldn't dare tell anyone . . .

Darwin watched the apparition skip away over the fields. It was quite astonishing. He'd never seen wasps swarm in such a manner.

A piece of paper fluttered to the ground. The curious clown must have dropped it.

Darwin picked it up and read, aloud, '"Return me, Hex". What does—?'

The afternoon dozed on. The grassy bank went back to its buzzing, humming, flowering busyness.

On the forlorn shore, a man appeared, hid two buckets behind a rock, and removed his false nose.

Rincewind scanned the landscape while extracting his hat from inside his shirt.

This was one of the most famous islands in the history of techno-mancy? It looked, frankly, rather dull.

He'd been expecting forests and streams and a riot of creatures. You couldn't move for vibrant, striving life on Mono Island, home of the God of Evolution. Everything wanted to leave. But this place had a skinflint look. You'd need to be tough to survive here. You'd have to fit in.

He couldn't see any giant tortoises, but there were a couple of large, empty shells.

Rincewind picked up a length of driftwood, baked by the sun into something like stone, and hurried up a narrow path.

Hex was good. The man Rincewind was after was striding ahead of him along the track.

'Mr Lawson, sir!'

The man turned.

'Yes? Are you from the *Beagle*?'

'Yessir. Heave ho, sir,' said Rincewind. Lawson stared at him.

'Why do you wear that hat with "Wizzard" written on it?'

Rincewind thought fast. Thank goodness Roundworld had some strange customs.

'Crossing the Line ceremony, sir,' he said. 'Took a fancy to it!'

'Oh, King Neptune and so forth,' said Lawson, backing away a little. 'Jolly good. How can I help you?'

'Just wanted to shake you by the hand and say how glad we all are that you're doing such a wonderful job out here, sir,' said Rincewind, pumping the man's unresisting arm vigorously.

'We . . . that's is very kind of you, Mr – what was that noise?'

'Sorry? Shiver my timber, by the way.'

'That . . . whistling noise . . . ' said Lawson, uncertainly.

'Probably one of the tortoises?' said Rincewind, helpfully.

'They hiss or – wasn't that a thump?' said Lawson. Behind him, a small cloud of dust rose above the bushes.

'Didn't hear one, yo ho,' said Rincewind, still shaking the hand. 'Well, don't let me keep you, sir.'

Lawson gave him the look of a man who feels has inadvertently fallen into dribbling company. The hat was clearly preying on his mind.

'Thank you, my man,' he said, pulling his hand away. 'Indeed, I must go.'

He headed away at some speed, which increased when he noticed Rincewind following him, and completely failed to notice what was, after all, just another small, rubble-filled hole among many. Rincewind spotted it, though, and after some effort pulled out a small, warm lump.

Something hissed, behind him.

Rincewind had ascertained that the only way a giant tortoise could go as fast as him was by falling over a cliff, and also that they were highly unlikely to savage a man to death. Still, he was ready.

He turned, stick upraised.

Something, a greyish something, something just transparent enough to show the landscape behind it in a dreary light, was hovering a few feet away. It looked like a monk's robe for a very small monk, and minus the monk. The empty hood was more worrying than almost anything that could have filled it. There were no eyes, there was no face, but there *was* nevertheless a stare, as malignant as razor-blade pants.

Other robe-shadows appeared around the shape and began drifting towards it. When they reached it they vanished, and the central shape became darker and, somehow, more present.

Rincewind didn't turn and run. There was no point in trying to run from Auditors; they were certainly faster than anything with legs. But that wasn't the reason. If it was time to run, he'd considered, no other calculations applied. He wouldn't even worry that his escape route was blocked by solid lava; most things could be overcome if you ran at them hard enough. There was, however, another reason. It had pink toes.

'Why meddle?' said the Auditor. The voice sounded windy and uncertain, as if the speaker was having to assemble the words by hand. 'Entropy will always triumph.'

'Is it true that you die if you have an emotion?' said Rincewind. The Auditor was quite dark now, which meant that it has assembled enough mass to move something quite heavy, like a human head.

'We do not have emotion,' said the Auditor. 'It is a human aberration. In you we detect the physical manifestation recognisable to us as fear.'

'You can't just kill people, you know,' said Rincewind. 'That's against the rules.'

'We believe there may be no rules here,' said the Auditor, moving forward.

'Wait, wait, wait!' said Rincewind, trying to back away into solid rock. 'You're saying you don't *know* what fear is, right?'

'We have no requirement to do so,' said the Auditor. 'Prepare to cease coherent function.'

'Turn around,' said Rincewind.

And a weakness of the Auditors is that they find a direct command hard to disobey, at least for a second or two. It turned, or, rather, flowed through itself to face the other way.

The lid of the Luggage closed with a 'clop' like the sound of a trout taking an unwary mayfly.

I wonder if it found out what fear really is, Rincewind thought.

But more grey shapes were distilling out of the air.

Now it was time to run.

EIGHTEEN
STEAM ENGINE TIME

 THERE WAS DARWIN, SITTING ON a bank, watching the bees, the wasps, the flowers . . . In the last paragraph of *The Origin* we find a beautiful and important passage that hints at afternoons of that kind:

It is interesting to contemplate an entangled bank, clothed with many plants of many kinds, with birds singing on the bushes, with various insects flitting about, and with worms crawling through the damp earth, and to reflect that these elaborately constructed forms, so different from each other, and dependent on each other in so complex a manner, have all been produced by laws acting around us.

Go ahead Paley, make my day.

All that wizardly effort to get him to write *The Origin*, not *The Ology*. It mattered to Darwin, of course, and it matters to those who chart the course of history. But, just as we can ask whether Lincoln's assassination really had much effect on subsequent events, so we can ask the same about Darwin's life's work. Would it really have mattered if the wizards had failed?

Metaphorical wizards, you appreciate. Yes, those happy coincidences that got Charles on board the *Beagle* and kept him there do look a tad suspicious, but *wizards*?

Let's ask the question in a more respectable way. How radical was Darwin's theory of natural selection, really? Did he have insights that no one before him had considered? Or did he just happen to be the person who caught the public eye, with an idea that had been float-ing around for some time? How much credit should he be given?

The same can be – and has been – asked of many 'revolutionary' scientific concepts. Robert Hooke got the idea of inverse square-law gravity before Newton did. Minkowski, Poincaré, and others worked out much of special relativity before Einstein did. Fractals were around, in some form, for at least a century before Benoît Mandelbrot energetically promoted them and they developed into a major branch of applied mathematics. The earliest sniff of chaos theory can be found in Poincaré's prize-winning memoir on the stability of the solar system in 1890, probably 75 years before the subject was perceived as 'taking off'.

How do scientific revolutions get started, and what decides who gets the credit? Is it talent? A flair for publicity? A lottery?

Part of the answer to these questions can be found in Robert Thurston's 1878 study of another important Victorian innovation, which Ponder Stibbons unerringly homed in on in Chapter 3. The book is *A History of the Growth of the Steam Engine*. The second paragraph says:

> History illustrates the very important truth: inventions are never, as great discoveries are seldom, the work of any one mind. Every great invention is really either an aggregation of minor inventions, or the final step in a progression. It is not a creation, but a growth as truly so as is that of the trees in the forest. The same invention is frequently brought out in several countries, and by several individuals, simulta-neously.

Thurston's topic reminds us of a common metaphor for this kind of apparently simultaneous invention: *steam engine time*. When it's steam engine time, suddenly everyone is making steam engines. When it's evolution time, everyone is inventing a theory of evolu-tion. When it's VCR time, everyone is making video cassette recorders. When it's Dotcom time, everyone is setting up Internet trading systems. And when it's Dotcom-going-bust time, all the Dotcoms are going bust.

There are times when human affairs really do seem to run on pre-constructed tracks. Some development becomes inevitable, and suddenly it's everywhere. Yet, just before that propitious moment, it wasn't inevitable at all, otherwise it would have happened already. 'Steam engine time' is a convenient metaphor for this curious process. The invention of the steam engine wasn't the first example, and it certainly wasn't the last, but it is one of the best known, and it's quite well documented.

Thurston distinguishes invention from discovery. He says that inventions are *never* the creation of a single individual, whereas great discoveries *seldom* are. However, the distinction isn't always clear-cut. Did ancient humans discover fire as a phenomenon of nature, or did they invent fire as a technology to keep predators away, light the cave, and cook food? The natural phenomenon surely came first, in the form of brush- or forest fires triggered by lightning, or possibly a droplet of water accidentally acting as a lens to concentrate the Sun's rays on to a piece of dry grass.*

However, that kind of 'discovery' doesn't *go* anywhere until someone finds a use for it. It was the idea of controlling fire that made the difference, and that seems more of an invention than a discovery. Except . . . you find out how to control fire by discovering that fires don't spread (so easily) across bare soil, that they can be spread very easily indeed by picking up a burning stick and dropping it into dry brushwood, or taking it home to the cave . . .

The inventive step, if there is such a thing, consists of putting together several independent discoveries so that what emerges has genuine novelty.

---

* Dry grass and drops of water are not commonly associated, but perhaps a damp elephant had just emerged from a river crossing on to dry savannah . . . Oh, invent your own explanation.

So inventions are often preceded by a series of discoveries. Similarly, discoveries are often preceded by inventions. The discovery of sunspots rested on the invention of the telescope, the discovery of amoebas and parameciums in pond water rested on the invention of the microscope. In short, invention and discovery are intimately entwined, and it's probably pointless to try to separate them. Moreover, the significant instances of both are much easier to spot in retrospect than they were at the time they first happened. Hindsight is a wondrous thing, but it does have the virtue of providing an explicit context for working out what did, or did not, matter. Hindsight lets us organise the remarkably messy process of invention/discovery, and tell convincing stories about it.

The problem is, most of those stories aren't true.

As children, many of us learned how the steam engine was invented. The young James Watt, aged about six, was watching a kettle boil, and he noticed that the pressure of the steam could lift the lid. In a classic 'eureka' moment, it dawned on him that a really big kettle could lift really heavy bits of metal, and the steam engine was born.

The original teller of this story was the French mathematician François Arago, author of one of the first biographies of Watt. For all we know, the story may be true, though it is more likely a 'lie-to-children', or educational aid,* like Newton's apple. Even if the young Watt was indeed suddenly inspired by a boiling kettle, he was by no means the first person to make the connection between steam and motive power. He wasn't even the first person to build a working steam engine. His claim to fame rests on something more complex, yet more significant. In Watt's hands, the steam engine became an effective and reliable tool. He didn't 'perfect' it – many smaller improvements were made after Watt – but he brought it into pretty much its final form.

*. See *The Science of Discworld*.

Watt wrote in 1774: 'The fire engine [= steam engine] that I have invented is now going, and answers much better than any other that has yet been made.' In conjunction with his business partner Matthew Boulton, Watt made himself the household name of the steam engine. And it has done his reputation no harm that, in the words of Thurston: 'Of the personal history of the earlier inventors and improvers of the steam-engine, very little is ascertained; but that of Watt has become well known.'

Was Darwin just another Watt? Did he get credit for evolution because he brought it into a polished, effective form? Is he famous because we happen to know so much about his personal history? Darwin was an obsessive record-keeper, he hardly threw away a single scrap of paper. Biographers were able to document his life in exceptional detail. It certainly did his reputation no harm that such a wealth of historical material was available.

In order to make comparisons, let's review the history of the steam engine, avoiding lies-to-children as much as we can. Then we'll look at Darwin's intellectual predecessors, and see whether a common pattern emerges. How does steam engine time work? What factors lead to a cultural explosion, as an apparently radical idea 'takes off' and the world changes for ever? Does the idea change the world, or does a changing world generate the idea?

Watt completed his first significant steam engine in 1768, and patented it in 1769. It was preceded by various prototypes. But the first recorded reference to steam as a source of motive power occurs in the civilisation of ancient Egypt, during the Late Kingdom when that country was under Roman rule. Around 150 BC (the date is very approximate) Hero of Alexandria wrote a manuscript *Spiritalia seu Pneumatica*. Only partial copies have survived to the present day, but from them we learn that the manuscript referred to dozens of steam-driven machines. We even know that several of them predated

Hero, because he tells us so; some were the previous work of the inventor Cestesibus, celebrated for the great number and variety of his ingenious pneumatic machines. So we can see the beginnings of steam engine time long ago, but initial progress was so quiet and slow that steam engine time itself was still far in the future.

One of Hero's devices was a hollow airtight altar, with the figure of a god or goddess on top, and a tube running through the figure. Unknown to the punters, the altar contains water. When a worshipper lights a fire on top of the altar, the water heats up and produces steam. The pressure of the steam drives some of the remaining liquid water up the pipe, and the god offers a libation. (As miracles go this one is quite effective, and distinctly more convincing than a statue of a cow that oozes milk or one of a saint that weeps.) Similar devices were commonplace from the 1960s to make tea at the bedside and pour it out automatically. They still exist today, but are harder to find.

Another of Hero's machines used the same principle to open a temple door when someone lit a fire on an altar. The device is quite complicated, and we describe it to show that these ancient machines went far beyond being mere toys. The altar and door are above ground, the machinery is concealed beneath. The altar is hollow, filled with air. A pipe runs vertically down from the altar into a metal sphere full of water, and a second inverted U-shaped pipe acts like a siphon, with one end inside the sphere and the other inside a bucket. The bucket hangs over a pulley, and ropes from the bucket wind round two vertical cylinders, in line with the hinges of the door and attached to the door's edge. They then run over a second pulley and terminate in a heavy weight which acts as a counterbalance. When a priest lights the fire, the air inside the altar expands, and the pressure drives water out of the sphere, through the siphon, and into the bucket. As the bucket descends under the weight of water, the ropes cause the cylinders to turn, opening the doors.

Then there's a fountain that operates when the sun's rays fall on it, and a steam boiler that makes a mechanical blackbird sing or blows a horn. Yet another device, often referred to as the world's first steam engine, boils water in a cauldron and uses the steam to turn a metal globe about a horizontal axis. The steam emerges from a series of bent pipes around the sphere's 'equator', at right angles to the axis.

In design, these machines weren't toys, but as far as their applications went, they might as well have been. Only the door-opener comes close to doing anything we would consider practical, although the priests probably found the ability to produce miracles on demand to be quite profitable, and that's practical enough for most businessmen today.

Looking back from the twenty-first century, it seems astonishing that it took steam engine time so long to gain proper momentum, with all these examples of steam power on public display all over the ancient world. Especially since there was plenty of demand for mechanical power, for the same reasons that finally gave birth to steam engine technology in the eighteenth century – pumping water, lifting heavy weights, mining, and transport. So we learn that it takes more than the mere ability to *make* steam engines, even in conjunction with a clear need for something of that kind, to kick-start steam engine time.

And so the steam engine bumbled along, never disappearing entirely, but never making any kind of breakthrough. In 1120 the church at Rheims had what looks suspiciously like a steam-powered organ. In 1571 Matthesius described a steam engine in a sermon. In 1519 the French academic Jacob Besson wrote about the production of steam and its mechanical uses. In 1543 the Spaniard Balso de Garay is reputed to have suggested the use of steam to power a ship. Leonardo da Vinci described a steam-gun that could throw a heavy metal ball. In 1606 Florence Rivault, gentleman of the bedchamber to Henry IV, discovered that a metal bombshell would explode if it was filled with water and heated. In 1615, Salomon de Caius, an engineer under Louis XIII, wrote about a machine that used steam to raise water. In 1629 . . . but you get the idea. It went on like that, with person after person reinventing the steam engine, until 1663.

In that year Edward Somerset, Marquis of Worcester, not only invented a steam-powered machine for raising water: he got it built, and installed, two years later, at Vauxhall – now part of London, but then just outside it. This was probably the first genuine application of steam power to a serious practical problem. No drawing of the machine exists, but its general form has been inferred from grooves, still surviving, in the walls of Raglan Castle, where it was installed. Worcester planned to form a company to exploit his machine, but failed to raise the cash. His widow in her turn made the same attempt, with the same lack of success. So that's another necessary ingredient for steam engine time: money.

In some ways, Worcester was the true creator of the steam engine, but he gets little credit, because he was just a tiny bit ahead of the wave. He does mark a moment at which the whole game changed, however: from this point on, people didn't just invent steam engines – they used them. By 1683, Sir Samuel Morland was building steam-powered pumps for Louis XIV, and his book of that year reveals a deep familiarity with the properties of steam and the associated mechanisms. The idea of the steam engine had now arrived, along with a few of the things themselves, earning their living by performing useful tasks. But it still wasn't steam engine time.

Now, however, the momentum began to grow rapidly, and what gave it a really big push was mining. Mines, for coal or minerals, had been around for millennia, but by the start of the eighteenth century they were becoming so big, and so deep, that they ran into what quickly became the miner's greatest enemy: water.

The deeper you try to dig mines, the more likely they are to become flooded, because they are more likely to run into underlying reservoirs of water, or cracks that lead to such reservoirs, or just cracks down which water from above can flow. Traditional methods of removing water were no longer successful, and something radically different was needed. The steam engine filled the gap neatly. Two people, above all, made it possible to build suitable machinery: Dennis Papin and Thomas Savery.

Papin trained in mathematics under the Jesuits at Blois, and in medicine in Paris, where he settled in 1672. He joined the laboratory of Robert Boyle, who would nowadays be called an experimental physicist. Boyle was working on pneumatics, the behaviour of gases –'Boyle's law', relating the pressure and volume of a gas at constant temperature, continues to be taught to this day. Papin invented the double air pump and the air gun, and then he invented the Digester. This is best described as a pressure-cooker, which is a saucepan with thick walls and a thick lid, held on securely so that water inside boils to form high-pressure steam. Food contained in the pan cooks very quickly.

The cookery aspect doesn't affect our story, but one bit of technology does. To avoid explosions, Papin added a safety valve, a feature replicated in the sixties domestic version, and an important invention because early involvement with steam engines was dangerous at the best of times. The idea probably originated earlier, but Papin gets the credit for using it to control steam pressure. In 1687 he moved to the University of Marburg, where he invented the first mechanical steam engine and the first piston engine. Throughout his career, he carried out innumerable experiments with steam-related apparatus, and introduced many significant pieces of gadgetry.

Steam engine time was hotting up. Savery, who also trained in mathematics, brought it to the boil. In 1698 he patented the first steam-powered pump that was actually used to clear mines of unwanted water – in this case, the deep mines of Cornwall. He sent a working model to the Royal Society, and later showed a model 'fire engine', as the machines were then confusingly called, to William III. The King granted him a patent:

A grant to Thomas Savery of the sole exercise of a new invention by him invented, for raising of water, and occasioning motion to all sorts of mill works, by the important force of fire, which will be of great use in draining mines, serving towns with water, and for the working of all sorts of mills, when they have not the benefit of water nor constant winds; to hold for 14 years; with usual clauses.

Steam engine time was close at hand. What clinched it was that Savery was a born businessman. He didn't wait for the world to beat a path to his door: he *advertised*. He gave lectures at the Royal Society, some of which were published in its journals. He circulated a prospectus among mine-owners and managers. And the selling point, naturally, was profit. If you can open up deeper levels of your mine, you can extract more minerals and make more money out of the same mine and the same bit of land.

Two more major steps were needed before what Thurston calls the 'modern' steam engine – that of 125 years ago – became firmly established. The first was to move from specialised, single-purpose machines, to multi-purpose ones. The second was to improve the engine's efficiency.

The move to multi-purpose steam engines was made by Thomas Newcomen, a blacksmith by trade, who introduced a radical new kind of engine, the 'atmospheric steam engine'. Previous engines had effectively combined a steam-driven piston and a pump in the same apparatus. Newcomen separated the components, and threw in a separate boiler and a condenser to boot. The piston moves up and down like a 'nodding donkey', driving a rod, which can be attached to . . . anything you like. Another engineer who must be mentioned here was John Smeaton, who scaled Newcomen's design up to much larger size.

Now, finally, we come to James Watt. Whatever credit he deserves, it is clear that he stood on the shoulders of a number of giants. Even if he had been capable of inventing the steam engine on his own, the plain fact is that he didn't. His grandfather was a mathematician – there seem to be a lot of mathematicians in the history of the steam engine – and Watt inherited his abilities. He carried out lots of experiments, and he made quantitative measurements, a relatively new idea. He worked out how heat travelled through the materials of the engine, and how much coal it took to boil a given amount of water. And he realised that the key to an efficient steam engine was to control unnecessary heat loss. The worst loss occurred in the cylinder that powered the piston, which kept changing temperature. Watt realised that the cylinder should always be kept at the same temperature as the steam that entered it – but how could that be done? The answer, when he finally chanced upon it, was simple and elegant:

> I had gone to take a walk on a fine Sabbath afternoon. I had entered the Green by the gate at the foot of Charlotte Street, and had passed the old washing-house. I was thinking upon the engine at the time, and had gone as far as the herd's house, when the idea came into my mind that, as steam was an elastic body, it would rush into a vacuum, and, if a communication were made between the cylinder and an exhausted vessel, it would rush into it, and might be condensed there without cooling the cylinder . . . I had not walked farther than the Golf-house, when the whole thing was arranged in my mind.

Such an easy thing to come up with – don't cool the steam in the cylinder, cool it *somewhere else*. Yet it improved the machine's efficiency so much that within a few years the only steam engines that anyone even thought of installing were those of Watt and his financial partner Boulton. Boulton-and-Watt engines cornered the market. No really significant improvements were subsequently made to their design. Or, to be more accurate, later 'improvements' supplanted the steam engine with engines of a very different design, driven by coal and oil. The steam engine had evolved to the pinnacle of its existence, and what displaced it was, in effect, a new species of engine altogether.

In retrospect, steam engine time arrived around the period of Savery, when the ability to make practical machines coincided with a genuine need for them in an industry that could afford to pay for them and would make more profits as a result. Add to that a sound business mind, to notice the situation and exploit it, and a sense for publicity to raise money from investors and get the idea off the ground, and the steam engine went like a . . . train.

Ironically, before most people realised that steam engine time had arrived, it had gone again, and in the end there was only one winner. The rest of the competition fell by the wayside. And that is why Watt gets so much credit, and why, ultimately, he deserves it. But he also deserves credit for his systematic quantitative experiments, his focus on the theory behind the steam engine, and his development of the concept – not as its inventor.

Certainly not for watching a kettle as a kid.

The history of the introduction of the Boulton-and-Watt steam engine is essentially an evolutionary one: the fittest design survived, the less fit were superseded and vanished from the historical record. Which brings us to Darwin, and natural selection. The Victorian era was 'steam engine time' for evolution; Darwin was just one of many people who recognised the mutability of species. Does he deserve the credit he gets? Was he, like Watt, the person who brought the theory to its culmination? Or did he play a more innovative role?

In the introduction to *Origin*, Darwin mentions several of his predecessors. So he certainly wasn't trying to take credit for the ideas of others. Unless you subscribe to the rather Machiavellian school of thought that giving credit to others is just a sneaky way of damning them with faint praise. One predecessor that he does *not* mention is perhaps the most interesting of all – his own grandfather, Erasmus

Darwin. Perhaps Charles felt that Erasmus was a bit too nutty to mention, especially being a relative.

Erasmus knew James Watt, and may have helped him to promote his steam engine. They were both members of the Lunar Society, an organisation of Birmingham technocrats. Another was Josiah Wedgwood, Darwin's uncle Jos's grandfather and founder of the famous ceramics company. The 'Lunaticks' met once a month at the time of the full moon – not for pagan or mystic reasons, or because they were all werewolves, but because that way they could see their way easily as they rode home after a few drinks and a good meal.

Erasmus, a physician, could also turn a nifty hand to machinery, and he invented a new steering mechanism for carriages, a horizontal windmill to grind Josiah's pigments, and a machine that could speak the Lord's Prayer and the Ten Commandments. When the 1791 riots against 'philosophers' (scientists) and for 'Church and King' put paid to the Lunar Society, Erasmus was just putting the finishing touches to a book. Its title was *Zoonomia*, and it was about evolution.

Not, however, by Charles's mechanism of natural selection. Erasmus didn't really describe a mechanism. He just said that organisms could change. All plant and animal life, Erasmus thought, derived from living 'filaments'. They *had* to be able to change, otherwise they'd still be filaments. Aware of Lyell's Deep Time, Erasmus argued that:

In the great length of time, since the earth began to exist, perhaps millions of ages before the commencement of the history of mankind, would it be too bold to imagine, that all warm-blooded animals have arisen from one living filament, which the first great cause endowed with animality, with the power of acquiring new parts, attended by new propensities, directed by irritations, sensations, volitions, and associations; and thus possessing the faculty of continuing to improve by its own inherent activity, and of delivering down those improvements by generation to its posterity, world without end!

If this sounds Lamarckian, that's because it was. Jean-Baptiste Lamarck believed that creatures could inherit characteristics acquired by their ancestors – that if, say, a blacksmith acquired huge muscular arms by virtue of working for years at his forge, then his children would inherit similar arms, *without* having to do all that hard work. Insofar as Erasmus envisaged a mechanism for heredity, it was much like Lamarck's. That did not prevent him having some important insights, not all of them original. In particular, he saw humans as superior descendants of animals, not as a separate form of creation. His grandson felt the same, which is why he called his later book on human evolution *The Descent of Man.* All very proper and scientific. But Ridcully is right. 'Ascent' would have been better public relations.

Charles certainly read *Zoonomia*, during the holidays after his first year at Edinburgh University. He even wrote the word on the opening page of his 'B Notebook', the origin of *Origin*. So his grandfather's views must have influenced him, but probably only by affirming the possibility of species change.* The big difference was that from the very beginning, Charles was looking for a mechanism. He didn't want to point out *that* species could change – he wanted to know *how* they changed. And it is this that distinguishes him from nearly all of the competition.

The most serious competitor we have mentioned already: Wallace. Darwin acknowledges their joint discovery in the second paragraph of the introduction to *Origin*. But Darwin wrote an influential and controversial book, whereas Wallace wrote one short paper in a technical journal. Darwin took the theory much further, assembled much more evidence, and paid more attention to possible objections.

He prefaced *Origin* with 'An Historical Sketch' of views of the origin of species, and in particular their mutability. A footnote mentions a remarkable statement in Aristotle, who asked why the various parts

* What would have happened if *Darwin* had gone back in time and killed his own grandfather?

of the body fit together, so that, for example, the upper and lower teeth meet tidily, instead of grinding against each other. The ancient Greek philosopher anticipated natural selection:

> Wheresoever, therefore, all things together (that is all the parts of one whole) happened like as if they were made for the sake of something, these were preserved, having been appropriately constituted by an internal spontaneity; and whatsoever things were not thus constituted, perished, and still perish.

In other words: if by chance, or some unspecified process the components carried out some useful function, they would appear in later generations, but if they didn't, the creature that possessed them would not survive.

Aristotle would have made short shrift of Paley.

Next, Darwin tackles Lamarck, whose views date from 1801. Lamarck contended that species could descend from other species, mostly because close study shows endless tiny graduations and varieties within a species, so the boundaries between distinct species is much fuzzier than we usually think. But Darwin notes two flaws. One is the belief that acquired characteristics can be inherited – Darwin cites the giraffe's long neck as an example. The other is that Lamarck believed in 'progress' – a one-way ascent to higher and higher forms of organisation.

A long series of minor figures follows. Among them is one noteworthy but obscure fellow, Patrick Matthew. In 1831, he published a book about naval timber, in which the principle of natural selection was stated in an appendix. Naturalists failed to read the book, until Matthew drew attention to his anticipation of Darwin's central idea in the *Gardener's Chronicle* in 1860.

Now Darwin introduces a better-known forerunner, the *Vestiges of the Natural History of Creation*. This book was published anonymously in 1844 by Robert Chambers; it is clear that he was also its author. The medical schools of Edinburgh were awash with the realisation that entirely different animals have remarkably similar anatomies, suggesting a common origin and therefore the mutability of species. For example, the same basic arrangement of bones occurs in the human hand, the paw of a dog, the wing of a bird, and the fin of a whale. If each were a separate creation, God must have been running out of ideas.

Chambers was a socialite – he played golf – and he decided to make the scientific vision of life on Earth available to the common man. A born journalist, Chambers outlined not just the history of life, but that of the entire cosmos. And he filled the book with sly digs at 'those dogs of the clergy'. The book was an overnight sensation, and each successive edition slowly removed various blunders that had made the first edition easy to attack on scientific grounds. The vilified clergy thanked their God that the author had not begun with one of the later editions.

Darwin, who respected the Church, had to refer to *Vestiges*, but he also had to distance himself from it. In any case, he found it woefully incomplete. In his 'Historical Sketch', Darwin quoted from the tenth 'and much improved' edition, objecting that the anonymous author of *Vestiges* cannot account for the way organisms are adapted to their environments or lifestyles. He takes up the same point in his introduction, suggesting that the anonymous author would presumably say that:

> After a certain number of unknown generations, some bird had given birth to a woodpecker, and some plant to the misseltoe [sic], and that these had been produced perfect as we now see them; but this assumption seems to me to be no explanation, for it leaves the case of the coadaptations of organic beings to each other and to their physical conditions of life, untouched and unexplained.

More heavyweights follow, interspersed with lesser figures. The first heavyweight is Richard Owen, who was convinced that species could change, adding that to a zoologist the word 'creation' means 'a process he knows not what'. The next is Wallace. Darwin reviews his interactions with both, at some length. He also mentions Herbert Spencer, who considered the breeding of domesticated varieties of animals as evidence that species could change in the wild, without human intervention. Spencer later became a major populariser of Darwin's theories. He introduced the memorable phrase 'survival of the fittest', which unfortunately has caused more harm than good to the Darwinian cause, by promoting a rather simple-minded version of the theory.

An unexpected name is that of the Reverend Baden Powell, whose 1855 'Essays on the Unity of Worlds' states that the introduction of new species is a natural process, not a miracle. Credit for mutability of species is also given to Karl Ernst von Baer, Huxley, and Hooker.

Darwin was determined not to miss out anyone with a legitimate claim, and in all he lists more than twenty people who in various ways anticipated parts of this theory. He is absolutely explicit that he is not claiming credit for the idea that species can change, which was common currency in scientific circles – and, as Baden Powell shows, beyond. What Darwin is laying claim to is not the idea of evolution, but that of natural selection as an evolutionary mechanism.

So . . . we come full circle. Does an innovative idea change the world, or does a changing world generate the idea?

Yes.

It's complicity. Both of these things happen – not once, but over and over again, each progressively altering the other. Innovations redirect the course of human civilisation. New social directions encourage further innovation. The world of human ideas, and the world of things, recursively modify each other.

That is what happens to a planet when a species evolves that is not merely intelligent, but what we like to call *extelligent*. One that can store its cultural capital outside individual minds. Which lets that capital grow virtually without limit, and be accessible to almost anybody in any succeeding generation.

Extelligent species take new ideas and run with them. Before the ink was dry on *Origin*, biologists and laymen were already trying to test its ideas, shoot them down, push them further. If Darwin had written *Ology*, and if nobody else had written something like *Origin*, then Victorian extelligence would have been enfeebled, and perhaps the modern world would have taken longer to arrive.

But it was evolution time. *Somebody* would have written such a book, and soon. And in that alternat(iv)e world of if, he or she would have got the credit instead.

So it's only fair to give Darwin the credit in this world. Steam engine time notwithstanding.

NINETEEN

LIES TO DARWIN

 ARCHCHANCELLOR RIDCULLY'S MOUTH DROPPED OPEN.

'You mean killed?' he said.

+++ No +++, Hex wrote, +++ I mean vanished. Darwin disappears from Roundworld in 1850. This is a new development. That is to say, it has always happened, but has always happened only for the last two minutes +++

'I really hate time travel,' sighed the Dean.

'Kidnapped?' said Ponder, hurrying across the hall.

+++ Unknown. Phase space currently contains proto histories in which he reappears after a fraction of one second and others where he never reappears at all. Clarity must be restored to this new node +++

'And you only tell us this now?' said the Dean.

+++ It has only just happened +++

'But,' the Dean attempted, 'when you looked at this . . . history before, this wasn't happening!'

+++ Correct. But that was then then, this is then now. Something has been changed. I surmise that this is as a result of our activities. And, having happened, it has always happened, from the point of view of an observer in Roundworld +++

'It's like a play, Dean,' said Ponder Stibbons. 'The characters just see the act they're in. They don't see the scenery being shifted because that's not part of the play.'

+++ Despite being wrong in every important respect, that is a very good analogy +++ Hex wrote.

'Have you any idea where he is?' said Ridcully.

+++ No +++

'Well, don't just sit there, man, find him!'

Rincewind reappeared above the lawn, and rolled expertly when he hit the ground. Other wizards, nothing like so experienced at dealing with the vicissitudes of the world, lay about groaning or staggered around uncertainly.

'It wears off,' he said, as he stepped over them. 'You might throw up a bit at first. Other symptoms of rapid cross-dimensional travel are short-term memory loss, ringing in the ears, constipation, diarrhoea, hot flushes, confusion, bewilderment, a morbid dread of feet, disorientation, nose bleeds, ear twinges, grumbling of the spleen, widgeons, and short-term memory loss.'

'I think I'd like to . . . thing . . . end of your life thing . . . ' murmured a young wizard, crawling across the damp grass. Nearby, another wizard had pulled off his boots and was screaming at his toes.

Rincewind sighed and made a grab at an elderly wizard, who was staring around like a lost lamb. He was also soaking wet, having apparently also landed in the fountain.

He looked familiar. It was impossible to know all the wizards in UU, of course, but this one he had definitely seen before.

'Are you the Chair of Oblique Frogs?' he said.

The man blinked at him. 'I . . . don't know,' he said. 'Am I?'

'Or the Professor of Revolvings?' said Rincewind. 'I used to write down my name on a piece of paper before this sort of thing. That's always a help. You *look* a bit like the Professor of Revolvings.'

'Do I?' said the man.

This looked like a very bad case. 'Let's find you your pointy hat and some cocoa, shall we? You'll soon feel—'

The Luggage landed with a thump, raised itself on its legs, and trotted away. The possible Professor of Revolvings stared at it.

'That? Oh, it's just the Luggage,' said Rincewind. The man didn't move. 'Sapient pearwood, you know?' Rincewind carried on, watching him anxiously. 'It's very clever wood. You can't get the very clever wood any more, not around here.'

'It moves about?' said the possible professor.

'Oh, yes. Everywhere,' said Rincewind.

'I know of no plant life that moves about!'

'Really? I wish I didn't,' said Rincewind, fervently, gripping the man a little tighter. 'Come on, after a nice warm drink you'll—'

'I must examine it closely! I am aware, of course of the so-called Venus Fly—'

'Please don't!' Rincewind pleaded, pulling the man back. 'You cannot botanise the Luggage!'

The bewildered man looked around with a desperation that was shading into anger.

'Who are you, sir? Where is this place? Why are all these people wearing pointy hats? Is this Oxford? *What has happened to me!*'

A chilly feeling was creeping over Rincewind. Quite probably, he alone of all the wizards had read Ponder's briefings as they arrived by surly porter; it paid to know what you might have to run away from. One had included a picture of a man who looked as if he was evolving all by himself, an effect caused by the riot of facial hair. This man was not that man. Not yet. But Rincewind could see that he would be.

'Um,' he said, 'I think you should come and meet people.'

It seemed to the wizards that Mr Darwin took it all very well, after the initial and quite understandable screaming.

It helped that they told him quite a lot of lies. No one would like to be told that they came from a universe created quite by accident and, moreover, by the Dean. It could only cause bad feeling. If you were told you were meeting your maker, you'd want something better.

It was Ponder and Hex who solved that. Roundworld's history offered a lot of opportunities, after all.

'I didn't *feel* any lightning strike,' Darwin said, looking around the Uncommon Room.

'Ah, you wouldn't have done,' said Ponder. 'The whole force of it threw you here.'

'Another world . . . ' said Darwin. He looked at the wizards. 'And you are . . . magical practitioners . . .'

'Do have a little more sherry,' said the Chair of Indefinite Studies. The sherry glass in Darwin's hand filled up again.

'You *create* sherry?' he said, aghast.

'Oh no, that's done by grapes and sunshine and so on,' said Ridcully. 'My colleague just moved it from the decanter over there. It's a simple trick.'

'We're all very good at it,' said the Dean cheerfully.

'Magic is basically just movin' stuff around,' said Ridcully, but Darwin was looking past him. The Librarian had just knuckled into the room, wearing the old green robe he wore for important occasions or when he'd had a bath. He climbed into a chair and held up a glass; it filled instantly, and a banana dropped into it.

'That is *Pongo pongo!*' said Darwin, pointing a shaking finger. 'An ape!'

'Well done that man!' said Ridcully. 'You'd be amazed at how many people get that wrong! He's our Librarian. Very good at it, too. Now, Mr Darwin, there's a delicate matter we—'

'It's another vision, isn't it?' said Darwin. 'It's my health, I know it. I have been working too hard.' He tapped the chair. 'But this wood feels solid. This sherry is quite passable. But magic, I must tell you, does not exist!' Beside him, with a little gurgle, his glass refilled.

'Just one moment, sir, please,' said Ponder. 'Did you say *another* vision?'

Darwin put his head in his hands. 'I though it was an epiphany,' he groaned. 'I thought that God himself appeared unto me and explained His design. It made so much sense. I had relegated Him to the status of Prime Mover, but now I see that He is immanent in His creation, constantly imparting direction and meaning to it all . . . or,' he looked up, blinking, 'so I thought . . . '

The wizards stood frozen. Then, very carefully, Ridcully said: 'Divine visitation, eh? And when was this, exactly?'

'It would have been after breakfast,' moaned Darwin. 'It was raining, and then I saw this strange beetle on the window. The room filled up with beetles—'

He stopped, mouth open; a thin blue haze surrounded him.

Ridcully lowered his hand.

'Well, well,' he said. 'What about *that*, Mr Stibbons?'

Ponder was scrabbling desperately at the paper on his clipboard.

'I've no idea!' he said. 'Hex hasn't mentioned it!'

The Archchancellor grinned the grim little grin of someone sensing that the game, at last, was afoot.

'Mono Island, remember?' said Ridcully, while Darwin stared blankly at nothing. 'A god with a thing about beetles?'

'I'd rather forget,' Ponder shuddered. 'But, but . . . no, it couldn't be *him*. How could the God of Evolution get into Roundworld?'

'Same way the Auditors did?' said Ridcully. 'All the spacetime continuumuum stuff we're doing, who's to say we aren't leaving a few doors ajar? Well, we can't let the barmy old boy run around there! You and Rincewind, meet me in the Great Hall in one hour!'

Ponder remembered the God of Evolution, who had been so proud of developing a creature even better fitted to survive than mankind. It had been a cockroach.

'We should go right away,' he said, firmly.

'Why? We can move in time!' said Ridcully. 'The hour, Mr Stibbons, is for you to come up with some way to kill Auditors!'

'They're indestructible, sir!'

'All right – ninety minutes!'

TWENTY

THE
SECRETS OF LIFE

 THE DISCWORLD VERSION OF DARWIN'S vision may not be *quite* what Roundworld's historians of science like to tell us, but the two will have been done converged on to the same timeline if the wizards manage to have will defeated the Auditors, so we can concentrate on the after-effects of that convergence. In any case some features are common to both versions of Darwinian history, including apes, beetles, and parasitic wasps. By contemplating these organisms, and many others – especially those confounded barnacles, of course – Darwin was led to his grand synthesis.

Today, no area of biology remains unaffected by the discovery of evolution. The evidence that today's species evolved from different ones, and that this process still continues, is overwhelming. Very little modern biology would make sense without the over-arching framework of evolution. If Darwin were reincarnated today, he would recognise many of his ideas, perhaps slightly reformulated, in the conventional scientific wisdom. The big principle of natural selection would be one of them. But he would also observe debate, perhaps even controversy, about this fundamental pillar of his thinking. Not whether natural selection happens, not whether it drives much of evolution; but whether it is the *only* driving force.

He would also find many new layers of detail filling some of the gaps in his theories. The most important and far-reaching of these is

DNA, the magic molecule that carries genetic 'information', the physical form of heredity. Darwin was sure that organisms could pass on their characteristics to their offspring, but he had no idea how this process was implemented, and what physical form it took. Today we are so familiar with the role of genes, and their chemical structure, that any discussion of evolution is likely to focus mainly on DNA chemistry. The role of natural selection, indeed the role of organisms, has been downgraded: the molecule has triumphed.

We want to convince you that it won't stay that way.

Evolution by natural selection, the great advance that Darwin and Wallace brought to public attention, is nowadays considered to be 'obvious' by scientists of most persuasions and by most non-specialists outside the US Bible Belt. This consensus has arisen partly because of a general perception that biology is 'easy', it isn't a real, hard-to-understand science like chemistry or physics, and most people think that they know enough about it by a kind of osmosis from the general folk information. This assumption showed up amusingly at the Cheltenham Science Festival in 2001, when the Astronomer Royal Sir Martin Rees and two other eminent astronomers gave talks on 'Life Out There'.

The talks were sensible and interesting, but they made no contact with real modern biology. They were based on the kind of biology that is currently taught in schools, most of which is about thirty years out of date. Like almost everything in school science, because it takes at least that long for ideas to 'trickle down' from the research frontiers to the classroom. Most 'modern mathematics' is at least 150 years old, so thirty-year-old biology is pretty good. But it's not what you should base your thinking on when discussing cutting-edge science.

Jack, in the audience, asked: 'What would you think of three biologists discussing the physics of the black hole at the centre of the galaxy?' The audience applauded, seeing the point, but it took a couple of minutes for the scientists on the platform to understand the symmetry. They were then as contrite as they could be without losing their dignity.

This kind of thing happens a lot, because we are all so familiar with evolution that we think we understand it. We devote the rest of this section to a reasonable account of what the average person thinks about evolution. It goes like this.

Once upon a time there was a little warm pond full of chemicals, and they messed about a bit and came up with an amoeba. The amoeba's progeny multiplied (because it was a good amoeba) and some of them had more babies (something funny here . . .) and some had fewer, and some of them invented sex and had a much better time after that. Because biological copying wasn't very good in those days, all of their progeny were different from each other, carrying various copying mistakes called *mutations*.

Nearly all mutations were bad, on the principle that putting a bullet randomly through a piece of complex machinery is unlikely to improve its performance, but a few were good. Animals with good mutations had many more babies, and those had the good mutation too, so they thrived and bred. Their progeny carried the good mutation into the future. However, many more bad mutations accumulated, so natural selection killed those off. Luckily, another new mutation appeared, which made a new character for a new species (better eyes, or swimming fins, or scales), which was altogether better and took over.

These later species were fishes, and one of them came out on land, growing legs and lungs to do so. From these first amphibians arose the reptiles, especially the dinosaurs (while the unadventurous fishes were presumably just messing about in the sea for millions of years, waiting to be fish and chips). There were some small, obscure mammals, who survived by coming out at night and eating dinosaur eggs. When the dinosaurs died, the mammals took over the planet, and some evolved into monkeys, then apes, then Stone Age people.

Then evolution stopped, with amoebas in ponds content to remain amoebas and not wanting to be fishes, fishes not wanting to be dinosaurs but just living their little fishy lives, the dinosaurs wiped out by a meteorite. The monkeys and apes, having seen what it was like to be at the peak of evolution, are now just slowly dying out – except in zoos, where they are kept to show us what our progenitors used to be like. Humans now occupy the top branch of the tree of life: since we are perfect, there's nowhere for evolution to go any more, which is *why* it has stopped.

If pressed for more detail, we dredge up various things we've learned, mostly from newspapers, about things called *genes*. Genes are made from a molecule called DNA, which takes the form of a double helix and contains a kind of code. The code specifies how to make that kind of organism, so human DNA contains the information needed to make a human, whereas cat DNA contains the information for a cat, and so on. Because the DNA helix is double, it can be split apart, and the separate parts can easily be copied, which is how living creatures reproduce. DNA is the molecule of life, and without it, life would not exist. Mutations are mistakes in the DNA copying process – typos in the messages of life.

Your genes specify everything about you – whether you'll be homosexual or heterosexual, what kinds of diseases you will be susceptible to, how long you will live . . . even what make of car you will prefer. Now that science has sequenced the human genome, the DNA sequence for a person, we know all of the information required to make a human, so we know everything there is to know about how human beings work.

Some of us will be able to add that most DNA isn't in the form of genes, but is just 'junk' left over from some distant part of our evolutionary history. The junk gets a free ride on the reproductive roller-coaster, and it survives because it is 'selfish' and doesn't care what happens to anything except itself.

Here ends the folk view of evolution. We've parodied it a little, but not by as much as you might hope. The first part is a lie-to-children about natural selection; the second part is uncomfortably close to 'neo-Darwinism', which for most of the past 50 years has been the accepted intellectual heir to *The Origin*. Darwin told us *what* happens in evolution; neo-Darwinism tells us *how* it happens, and how it happens is DNA.

There's no question that DNA is central to life on Earth. But virtually every month, new discoveries are being made that profoundly change our view of evolution, genetics, and the growth and diversification of living creatures. This is a vast topic, and the best we can do here is to show you a few significant discoveries and explain why they are significant.

Just as physics replaced Newton by Einstein, there has been a major revolution in the basic tenets of biology, so we now have a different, more universal view of what drives evolution. The 'folk' evolutionary viewpoint: 'I've got this new mutation. I have become a new kind of creature. Is it going to do me any good?' is not the way modern biologists think.

There are many things wrong with our folk-evolution story. In fact we've deliberately constructed it so that every single detail is wrong. However, it's not very different from many accounts in popular science books and television programmes. It assumes that primitive animals alive today are our ancestors, when they are our cousins. It assumes that we 'came from' apes, when of course the ape-like ancestor of man is the same creature as the man-like ancestor of modern apes. More seriously, it assumes that mutations in the genetic material, the changes that natural selection has to work on – indeed, to select among – are checked out as soon as they appear, and labelled 'bad' (the organism dies, or at least fails to breed) or 'good' (the animal contributes its progeny to the future).

Until the early 1960s, that was what most biologists thought too. Indeed, two very famous biologists, J.B.S. Haldane and Sir Ronald Fisher, produced important papers in the mid-1950s espousing just that view. In a population of about 1000 organisms, they believed, only about a third of the breeding population could be 'lost' to bad gene variants, or could be ousted by organisms carrying better versions, without the population moving towards extinction. They calculated that only about ten genes could have variants (known as 'alleles') that were increasing or decreasing as proportions of the population. Perhaps twenty genes might be changing in this way if they were not very different in 'fitness' from the regular alleles. This picture of the population implied that almost all organisms in a given species must have pretty much the same genetic make-up, except for a few which carried the good alleles coming in, and winning, or the bad alleles on the way out.* These exceptions were mutants, famously and stupidly portrayed in many SF films.

However, in the early 1960s Richard Lewontin's group exploited a new way to investigate the genetics of wild (or indeed any) organisms. They looked at how many versions of common proteins they could find in the blood, or in cell extracts. If there was just one version, the organism had received the same allele from both of its parents: the technical term here is 'homozygous'. If there were two versions, it had received different ones from each parent, and so was 'heterozygous'.

What they found was totally incompatible with the Fisher–Haldane picture.

They found, and this has been amply confirmed in thousands of wild populations since, that in most organisms, about *ten per cent* of

---

* They made exceptions for manifestly 'unimportant' but very diverse sets of alleles like blood groups, but in those cases it didn't seem to matter much which kind you had.

genes are heterozygous. We now know, thanks to the Human Genome Project, that human beings have about 34,000 genes. So about 3400 are heterozygous, in any individual, instead of the ten or so predicted by Haldane and Fisher.

Furthermore, if many different organisms are sampled, it turns out that about one-third of all genes have variant alleles. Some are rare, but many of them occur in more than one per cent of the population.

There is no way that this real-world picture of the genetic structure of populations can be reconciled with the classical view of population genetics. Nearly all current natural selection must be discriminating between different combinations of *ancient* mutations. It's not a matter of a new mutation arriving and the result being immediately subjected to selection: instead, that mutation must typically hang around, for millions of years, until eventually it ends up playing a role that makes enough of a difference for natural selection to notice, and react.

With hindsight, it is now obvious that all currently existing breeds of dog must have been 'available' – in the sense that the necessary alleles already existed, somewhere in the population – in the original domesticated wolves. There simply hasn't been *time* to accumulate the necessary mutations purely in modern dogs. Darwin knew about the amount of cryptic and overt variation in pigeons, too. But his successors, hot on the trail of the molecular basis of life, forgot about wolves and pigeons. They pretty much forgot about cells. DNA was complicated enough: cell biology was impossible, and as for understanding an *organism* . . .

Lewontin's discovery was a significant turning point in our understanding of heredity and evolution. It was at least as radical as the much better publicised revolution that replaced Newton's physics with Einstein's, and it was arguably more important. We will see that in the last year or so there has been another, even more radical, revision of our thinking about the control of cell biology and development by the genes. The whole dogma about DNA, messenger RNA, and proteins has been given a reality check, and science's internal 'auditors' have rendered it as archaic as Fisher's population genetics.

It is commonly assumed – not only by the average television producer of pop science half-hours, but also by most popular science book authors – that now we know about DNA, the 'secret of life', evolution and its mechanisms are an open book. Soon after the discovery of DNA's structure and mechanism of replication by James Watson and Francis Crick, in the late 1950s, the media – and biology textbooks at all levels – were beginning to refer to it as the 'Blueprint for Life'. Many books, culminating with Dawkins's *The Selfish Gene* in the 1970s, promoted the view that by knowing about the mechanism of heredity, we had found the key to all of the important puzzles of biology and medicine, especially evolution.

There was soon to be a major tragedy, resulting from a medical application of that mistaken view. The sedative thalidomide was increasingly being prescribed, and bought over the counter, to treat nausea and other minor discomforts of the early weeks of pregnancy. Only later was it discovered that in a small proportion of cases, thalidomide could cause a type of birth defect known as phocomelia, in which arms and legs are replaced by partially developed versions that resemble a seal's flippers.

It took a while for anyone to notice, partly because few general practitioners had experience of phocomelia before 1957. In fact, very few of them had ever seen a case at all, but after 1957 they began to see two or three in a year. A second reason was that it was very difficult to tie this defect to a particular potion or treatment: pregnant women famously take a great variety of dietary additives, and often they don't remember precisely what they've taken. Nevertheless, by 1961 some medical detective work had tied the spate of phocomelia down to thalidomide.

American doctors congratulated themselves on having missed out on the pathology, because Frances Kelsey, a medical worker for the Food and Drug Administration, had expressed misgivings about the original animal testing of the drug. Her misgivings eventually turned out to have been unfounded, but they did save much suffering in the USA. She noticed that the drug had not been tested on pregnant animals, because at that time such tests were not required. Everyone knew that the embryo has its own blueprint for development, quite separate from that of the mother. However, embryologists trained in biology departments, as distinct from medical embryologists, knew about the work of Cecil Stockard, Edward Conklin, and other embryologists of the 1920s. They had shown that many common chemicals could caused monstrous developmental defects. For instance, lithium salts easily induced cyclopia, a single central eye, in fish embryos. These alternative developmental paths, induced by chemical changes, have taught us a lot about the biological development of organisms, and how it is controlled.

They have also taught us that an organism's development is *not* rigidly determined by the DNA of its cells. Environmental insults can push the course of development along pathological paths. In addition, the genetics of organisms, particularly wild organisms, are usually organised so that 'normal' development happens *despite* a variety of environmental insults, and even despite changes in some of the genes. This so-called 'canalised' development is very important for evolutionary processes, because there are always temperature variations, chemical imbalances and assaults, parasitic bacteria and viruses; the growing organism must be 'buffered' against these variations. It must have versatile developmental paths to ensure that the 'same' well-adapted creature is produced, whatever the environment is doing. Within reasonable limits, at any rate.

There are many developmental tactics and strategies that help to accomplish this. They range from simple tricks like the HSP90 protein to the very clever mammalian trade-off.

HSP stands for 'heat shock protein'. There are about 30 of these proteins, and they are produced in most cells in response to a sudden, not very severe, change of temperature. A different array of proteins is produced in response to other shocks; this one is called HSP90 because of where it sits in a much longer list of cell proteins. HSP90, like most HSPs, is a chaperonin: its job is to hug other proteins during their construction, so that when the long line of amino acids folds up it achieves the 'right' shape. HSP90 is very good at making the 'right' shape – even if the gene that specifies the chaperoned protein has accumulated a lot of mutations. So the resulting organism doesn't 'notice' the mutations; the protein is 'normal' and the organism looks and behaves just like its ancestral form.

However, if there's a heat shock or other emergency during development, HSP90 is diverted from its role as chaperonin, and other less powerful chaperonins permit the mutational differences to be expressed in most of the progeny. The effect this has on evolution is to keep the organisms much the same until there's an environmental stress, when suddenly, in one generation, lots of previously hidden, but hereditable, variation appears.

Most books that describe evolution seem to assume that every time there's a mutation, the environment promptly gets to judge it good or bad . . . but one little trick, HSP90, which is present in most animals and many bacteria, makes nonsense of that assertion. And from Lewontin's discovery that a third of genes have common variants in wild populations, and that all organisms carry lots of them, it is clear that ancient mutations are continually being tested in different modern combinations, while the potential effects of more recent mutations are being cloaked by HSP90 and its ilk.

The trick employed by mammals is much more complex and far-reaching. They reorganised their genes, and got rid of a lot of genetic complication that their amphibian ancestors relied on, by adopting a new and more controlled developmental strategy.

Most frogs and fishes, whose eggs usually encounter great differences and changes of temperature during each embryology, ensure that the 'same' larva, and then adult, results. Think of frog spawn in a frozen English pond, warming up to 35°C during the day while the delicate early development proceeds; then the little hatchling tadpoles have to endure these temperature changes. Now think of the frogs that so few of the tadpoles become.

Most chemical reactions, including many biochemical ones, happen at different rates if the temperature is different. You only get a frog if all the different developmental processes fit together effectively, and timing is crucial. So how does frog development work at all, given that the environment is changing so quickly and repeatedly?

The answer is that the frog genome 'contains' many different contingency plans, for many different environmental scenarios. There are many different versions of each of the enzymes and other proteins that frog development requires. All of them are put into the egg while it is in mother frog's ovary. There are perhaps as many as ten versions of each, appropriate to different temperatures (fast enzymes for low temperatures, sluggish ones for higher temperatures, to keep the duration of development much the same*), and they have 'labels' on the packages that make them, so the embryo can choose which one to use according to its temperature. Animals whose development must be buffered in this way use a lot of their genetic programme to set up contingency plans for many other variables, in addition to temperature.

The mammals cleverly avoided all of this faffing around, by making their females thermostatically controlled – 'warm-blooded'. What

---

* That's very important for a few species. Zebra-fish eggs in the wild must hatch in just under 72 hours, because they're laid just before dawn and must hide before the third dawn when predators could see them.

counts is not the warmth of the blood, but the system that maintains it at a constant temperature. The beautifully controlled uterus keeps all kinds of other variables away from the embryos, too, from poisons to predators. It probably 'costs' much less in DNA programming to adopt this strategy, too.

This trick, evolved by the mammals, carries an important message. To ask how much information passes across the generations in the DNA blueprint, as textbooks and sophisticated research manuals often do, is to miss the point. How the genes and proteins are *used* is far more important, and far more interesting, than how many genes or proteins there are in a given creature. Lungfishes and some salamanders, even some amoebas, have more than fifty times as much DNA as we mammals do. What does this say about how complex these creatures are, compared to us?

Absolutely nothing.

Tricks like HSP90, and strategies like warm-bloodedness and keeping development inside the mother, mean that bean-counting of DNA 'information' is beside the point. What counts is what the DNA means, not how big it is. And meaning depends on context, as well as content: you can't regulate the temperature of a uterus unless your context (that is, mother) provides one.

The simple-minded 'mutation' viewpoint, allied to trendy interpretations of DNA function in terms of 'information theory', is often allied with ignorance of biology in other areas. One example is radiation biology and simple ecology as seen by 'conservation activists'. Some of these volunteers found five-legged frogs and other 'monsters' downwind of the Chernobyl site, years after the nuclear accident but while radiation levels were still noticeably high. They claimed that the monsters were mutants, caused by the radiation. Other workers, however, then found just as many supposed mutants upwind of the reactor site.

It turned out that the best explanation had nothing to do with mutant frogs. It was the *absence* of their usual predators, owls and hawks and snakes, because there were so many humans trudging about. *Rana palustris* tadpoles from Chernobyl produced no more of these pathologies than did other frogspawn samples from ponds some tens of kilometres away that had not been subjected to radiation, when a high percentage of both was allowed to survive. Usually, in British *Rana temporaria* frogs, it is very difficult to achieve ten per cent normal adults, or even ones that are viable in the laboratory, but they don't produce extra limbs as *palustris* does. It is normally the case, of course, that a female frog's lifetime production of some 10,000 eggs results in a few highly selected, and therefore 'normal', survivors, and on average just two breeders. But conservationists don't like thinking about this reproductive arithmetic, with all those deaths.

Here is another issue, again chosen from the thalidomide literature, that demonstrates how talk of Lamarckism, or of 'mutations', misses the point.

Some of the children affected by thalidomide have married each other, and several of these pairings have produced phocomelic children. The obvious deduction, from the folk-DNA point of view, is that the DNA of the first generation must have been altered, so that it produced the same effect in the next generation. In fact, this effect looks, at first glance, like Lamarckism: the inheritance of acquired characters. Indeed, it seems a classic demonstration of such inheritance, as convincing as if cutting off terriers' tails resulted in puppies being born with short tails. However, it is actually a lesson in not attempting to explain things 'at first glance', like the conservationists did with the abnormal frogs.

It is very tempting to do just that, when the idea of heredity in your mind is that one gene leads to one character, so if you've got the character you've got the gene, and vice versa. Figures from the epidemiological literature suggest that in the space of a few years either side of 1960, about 4 million women took thalidomide at the critical time during gestation. Of those, about 15,000–18,000 foetuses were damaged; 12,000 came to birth with defects, and about 8,000 survived their first year. That is to say, the natural course of development *selected* just 1 in 500 who showed adverse effects. The proportion of children born with no detectable defect was much, much higher. And that fact changes our view of the likely reason for the children of two thalidomide parents to suffer from phocomelia, for the following reason.

Conrad Waddington demonstrated a phenomenon called 'genetic assimilation'. He started with a genetically diverse population of wild fruit flies, and found that about one in 15,000 of their pupae, when warmed, produced a fly with no cross-vein in its wing. These 'cross-veinless' flies looked just like some very rare mutant flies that turned up occasionally in the wild, just as occasional genetically phocomelic children turned up before thalidomide. By breeding from the flies that responded to the treatment, Waddington selected for a lower and lower threshold of response. In a few tens of generations, he had selected flies that bred true for the cross-veinless trait, exhibiting it regularly *without* anyone warming the pupae. This may look like Lamarckian inheritance, but it's not. It's genetic assimilation. The experiments were selecting flies that had no cross-vein at lower and lower temperature thresholds. Eventually, they selected flies that had no cross-vein at 'normal' temperatures.

Similarly, genetic assimilation provides a much better explanation than Lamarckism for the phocomelic children of thalidomide-modified parents. We have *selected*, from some 4 million foetuses, those that respond to thalidomide with phocomelia. It is not surprising that when they marry each other, they produce a few progeny whose threshold is very low – below zero in fact. They are so liable to produce phocomelia that they do it without thalidomide, just as Waddington's flies came to produce cross-veinlessness without warming the pupae.

*

One of the things that really worried Darwin was the existence of parasitic wasps – a fact that has influenced our Discworld tale, but has gone unremarked until now in the scientific commentaries. Parasitic wasps lay their eggs in other insects' larvae, so that as the wasp eggs grow into wasp larvae, they eat their hosts. Darwin could see how this might have happened on evolutionary grounds, but it seemed to him to be rather immoral. He was aware that wasps don't have a sense of morality, but he saw it as some kind of flaw on the part of the wasps' creator. If God designed each species on Earth, for a special purpose – which is what most people believed at the time – then God had deliberately designed parasitic wasps, whose purpose was to eat other species of insect, also designed by God. To be so eaten, presumably.

Darwin was fascinated by such wasps, ever since he first encountered them in Botafogo Bay, Brazil. He eventually satisfied himself – though not his successors – that God had found it necessary to permit the existence and evolution of parasitic wasps *in order to get to humans*. This is what the quote at the end of Chapter 10 alludes to. That particular explanation has fallen out of favour among biologists, along with all theist interpretations. Parasitic wasps exist because there is something for them to parasitise – so why not? Indeed, parasitic wasps play a major role in controlling many other insect populations: nearly one-third of all of the insect populations that humans like to label 'pests' are kept at bay in this manner. Maybe they *were* created in order for humans to be possible . . . At any rate, the wasps that so puzzled Darwin still have much to tell us, and the latest discovery about them threatens to overturn several cherished beliefs.

Strictly, the discovery is not so much about the wasps, as about some viruses that infect them . . . or are symbiotic with them. They are called polydnaviruses.

When mother wasp injects her eggs into some unsuspecting larva, such as a caterpillar, she also injects a solid dose of viruses, among them said polydnaviruses. The caterpillar not only gets a parasite, it gets an infection. The virus's genes produce proteins that interfere with the caterpillar's own immune system, stopping it reacting to the parasite and, perhaps, rejecting it. So the wasp larvae munch merrily away on the caterpillar, and in the fullness of time they develop into adult wasps.

Now, any self-respecting adult parasitic wasp obviously needs its own complement of polydnaviruses. Where does it get them? From the caterpillar that it fed on. And it gets them (just as mother did) not as a separate infective 'organism', but as what is called a provirus: a DNA sequence that has been integrated into the wasp's own genome.

Many genomes, probably most if not all, include various bits of viruses in this way. Our own certainly does. Transport of DNA by viruses seems to have been an important feature of evolution.

In 2004 a team headed by Eric Espagne worked out the DNA sequence of a polydnavirus – as one does – and what they found was dramatically different from what anyone had expected. Typical virus genomes are very different from those of 'eukaryotes' – organisms whose cells have a nucleus, which includes most multi-cellular creatures and many single-celled ones, but not bacteria. The DNA sequences of most eukaryote genes consist of 'exons', short sequences that collectively code for proteins, separated by other sequences called introns, which get snipped out when the code is turned into the appropriate protein. Viral genes are relatively simple, and typically they do not contain introns. They consist of connected code sequences that specify proteins. This particular polydnavirus genome, in contrast, does contain introns, quite a lot of them. The genome is complex, and looks much more like a eukaryote genome than a virus genome. The authors conclude that polydnavirus genomes constitute 'biological weapons directed by the wasps against their hosts'. So they look more like the enemy's genome than that of an ordinary virus.

*

Numerous examples, old and new, disprove every aspect of the folk version of evolution and DNA. We end with one that looks especially important, discovered very recently, and whose significance is just becoming seriously apparent to the biological community. It is probably the most severe shock that cell biology has received since the discovery of DNA and the wonderful 'central dogma': DNA specifies messenger-RNA which specifies proteins. The discovery was not made through some big, highly publicised research programme like the human genome project. It was made by someone who wondered why his petunias had gone stripy. When all the world is chasing 'the' human genome, it's not easy to get research grants to work on stripy petunias. But what the petunias revealed is probably going to be far more important for medicine than the entire human genome project.

Because proteins *are* the structure of living creatures, and because as enzymes they control the processes of life, it has seemed obvious that DNA controls life, that we can 'map' DNA code on to all the important living functions. We could assign a function to each protein, so we could assume that the DNA that coded for that protein was ultimately or fundamentally responsible for the corresponding function. Dawkins's early books reinforced the idea of one gene, one protein, one function (although he carefully warned his readers that he didn't want to give that impression), and this encouraged such media exaggerations as calling the human genome the Book of Life. And the 'selfish gene' image made it entirely credible that huge stretches of the genome were present for solely selfish reasons – that is, for no reason related to the organism concerned.

Biologists employed – as so many now are – in the biotechnology industries serving agriculture, pharmacy, medicine, even some engineering projects (we don't mean just 'genetic engineering' but making better motor oils), all subscribe to the central dogma, with a few minor modifications and exceptions. All of them have been informed that nearly all of the DNA in the human genome is 'junk', not coding for proteins, and that although some of it may be important for developmental processes or for controlling some of the 'real' genes, they really don't need to worry about it.

Admittedly, quite a lot of junk DNA seems to be transcribed into RNA, but these are just short lengths that sit about briefly in the cell fluids and don't need to be considered when you're doing important protein-making things with the real genes. Recall that the DNA sequences of real genes consist of a mosaic of 'exons' which code for proteins, separated by other sequences called introns. The introns have to be cut out of the RNA copies to get the 'real' protein-coding sequences, called messenger-RNAs, which lace into ribosomes like tapes into a tape player. Messenger-RNAs determine what proteins get made, and they have sequences on their ends that label them for making many copies of a protein or for destruction after only a couple of protein molecules.

Nobody worried much about those snipped-out introns, just bits of RNA drifting aimlessly around in the cell till they got broken up by enzymes. Now, they do worry. Writing in the October 2004 *Scientific American*, John Mattick reports that

The central dogma is woefully incomplete for describing the molecular biology of eukaryotes. Proteins do play a role in the regulation of eukaryotic gene expression, yet a hidden, parallel regulatory system consisting of RNA that acts directly on DNA, RNAs and proteins is also at work. This overlooked RNA signalling network may be what allows humans, for example, to achieve structural complexity far beyond anything seen in the unicellular world.

Petunias made that clear. In 1990 Richard Jorgensen and colleagues were trying to breed new varieties of petunias, with more interesting, brighter colours. An obvious approach was to engineer into the petunia genome some extra copies of the gene that coded for an enzyme involved in the production of pigment. More enzyme, more pigment, right?

Wrong.

Less pigment?

No, not exactly. What previously was a uniformly coloured petal became stripy. In some places the pigment was being produced, elsewhere it wasn't. This effect was so surprising that plant biologists tried to find out exactly why it was happening. And what they found was 'RNA interference'. Certain RNA sequences can shut down a gene, prevent it making protein. It happens in many other organisms, too. In fact, it is extremely widespread. And it suggests something extraordinarily important.

The big question in this area, asked many times and largely ignored, has always been: if introns (which occupy all but one-twentieth of a typical protein-coding gene) have no biological function, why are they there? It is easy to dismiss them as relics of some dim evolutionary past, no longer useful, lying around because natural selection can't get rid of things that are harmless. Even so, we can still wonder whether introns are present because they do have some useful function, one that we haven't yet worked out. And it's starting to look as if that may be the case.

For a start, introns are not that ancient. It now seems that they became incorporated into the human genome relatively recently. They are probably related to mobile genetic elements known as group II introns, which are a 'parasitic' form of DNA that can invade host genomes and then remove themselves when the DNA is expressed as RNA. Moreover, they now seem to have a role as 'signals' in the regulation of genetic processes. An intron may be relatively short, compared to the long protein-coding sequences that arise when the introns are snipped out, but a short signal has advantages and can do quite a lot. In effect, the introns may be genetic 'txt msgs' in the mobile phone of life. Short, cheap, and very effective. An RNA-based 'code', running parallel to the DNA double helix, can affect the activity of the cell very directly. An RNA sequence can act as a very specific, well-defined signal, directing RNA molecules to their targets in RNA or DNA.

The evidence for the existence of such a signalling system is reasonable, but not yet undeniable. If such a system does exist, it clearly has the potential to resolve many biological mysteries. A big puzzle about the human genome is that its 34,000 genes manage to encode over 100,000 proteins. Clearly 'one gene one protein' doesn't work. A hidden RNA signalling system could make one gene produce several proteins, depending on what the accompanying RNA signal specified. Another puzzle is the complexity of eukaryotes, especially the Cambrian explosion of 525 million years ago, when the range of terrestrial body-plans suddenly diversified out of all recognition; indeed, was more diverse than it is now. Perhaps the hypothetical RNA signalling system started to take off at that time. And it's widely known that the human and chimpanzee genomes are surprisingly similar (though the degree of similarity seems not to be 98 per cent as widely quoted even a few years ago). If our RNA signals are significantly different, that would be one way to explain why humans don't greatly resemble chimps.

At any rate, it very much looks as if all that 'junk' DNA in your genome is not junk at all. On the contrary, it may be a crucial part of what makes you human.

This lesson is driven home by those business associates of parasitic wasps, the symbiotic polydnaviruses, sneakily buried inside the wasps' own DNA. There is a message there about human evolution, and it's a very strange one.

Genome-sequencing may have been oversold as *the* answer to human diseases, but it's very good basic science. The activities of the sequencers have revealed that wasps are not the only organisms to have bits of viral DNA hanging around in their genomes. In fact, most creatures do, humans included. The human genome even contains one *complete* viral genome, and only one, called ERV-3 (**E**ndogenous **R**etro**V**irus). This may seem an evolutionary oddity, a bit of 'junk

DNA' that really is junk . . . but, actually, without it none of us would be here. It plays the absolutely crucial role of preventing rejection of the foetus by the mother. Mother's immune system 'ought to' recognise the tissue of a developing baby as 'foreign', and trigger actions that will get rid of it. By 'ought to' here we mean that this is what the immune system normally does for tissue that is not the mother's own.

Apparently, the ERV-3 protein closely resembles another one called p15E, which is part of a widespread defence system used by viruses to stop their hosts killing them off. The p15E protein stops lymphocytes, a key type of cell in the immune system, from responding to antigens, molecules that reveal the virus's foreign nature. At some stage during mammalian evolution, this defence system was stolen from the viruses and used to stop the female placenta responding to antigens that reveal the foreign nature of the foetus's *father*. Perhaps on the principle of being hung for a sheep as well as for a lamb, the human genome decided to go the whole hog* and steal the entire retroviral genome.

When evolution carried out the theft, however, it did not just dump its booty into the human DNA sequence unchanged. It threw in a couple of introns, too, splitting ERV-3 into several separate pieces. It's *complete*, but not connected. No matter: enzymes can easily snip out the introns when that bit of DNA is turned into protein. But no one knows why the introns are there. They might be an accidental intrusion. Or – pursuing the RNA interference idea – they might be much more significant. Those introns might be an important part of the genetic regulatory system, 'text messages' that let the placenta use ERV-3 without running the risk of setting the corresponding virus loose.

At any rate, whatever the introns are for, warm-bloodedness is not the only trick that mammalian evolution managed to find and exploit.

* Sorry about the proliferation of barnyard metaphors.

It also indulged in wholesale theft of a virus's genome, to stop mother's immune system booting out baby because it 'smelled' of father. And we also get another lesson that DNA isn't *selfish*. ERV-3 is present in the human genome, but not because it's a bit of junk that gets copied along with everything else and has remained because it does no harm. It's there because, in a very real sense, humans could not survive – could not even *reproduce* – without it.

TWENTY-ONE

NOUGAT SURPRISE

 THE ACTIVITY IN THE GREAT HALL was slowing down now. All along the rainbow lines of time, the nodes were closed. Or shut, or denoded, Rincewind thought. Whatever you did with nodes, anyway. There was a little cheer as the last glowing wizard symbol faded away, and a roar from outside as three wizards and a lot of tentacles landed in the fountain. Rincewind had been surprised about that, and then dismayed. Since it did *not* have his name on it, this meant that Ridcully had something worse in mind.

'Looks like I'm not needed now,' he said hopefully, just in case.

'Haha, professor, what a card you are and no mistake,' said the lobster next to him. 'The Archchancellor was very definite that we was to keep you here no matter what you said.'

'But I wasn't running away!'

'No, you was only inspecting the wall with the loose bricks in it,' said the lobster.* 'We quite understand. Lucky for you we caught you before you dropped over into the alley, eh? Could've done yerself a mischief.'

Rincewind sighed. The lobsters were always hard to outrun. They hunted in packs, appeared to share a common brain, and many years

---

* The University's proctors were known as lobsters because they went very red when hot and had a grip that was extremely hard to shake off. They were generally ex-army sergeants, had depths of cynicism unplumbable by any line, and were fuelled by beer.

of harrying erring students had given them a malignant street cunning that verged on the supernatural.

Some of the senior wizards swept in . . . There was an argument going on between Ridcully and the Dean.

'I don't see why I shouldn't.'

'Because you get too excited in the presence of combat, Dean. You run around making silly "hut, hut" noises,' said Ridcully. 'Remember why we had to stop those paintball afternoons? You didn't seem to be able to get the hang of the term "people on my side".'

'Yes, but this is—'

'It took us a week to get the Senior Wrangler looking halfway suitable for polite company – Ah, Rincewind. Still with us, I see. Good man. Very well, Mr Stibbons: report!'

Ponder coughed.

'Hex confirms that, er, that our recent activities may have left camiloops between our word and Roundworld, sir. That is to say, residual connections that may be used deliberately or inadvertently from either side. Magic doors, in fact, drifting without anchor. These will evaporate within a matter of days. Um . . .'

'I don't want to hear "um", Mr Stibbons. "Um" is not a word we entertain here.'

'Well, the fact is that since the camiloops are spread over centuries, the Auditors may very well have been in Roundworld for some time. We have no way of knowing for how long. Hex, um, sorry, does report some circumstantial evidence that humans are dimly aware of their malign meddling, albeit at a very mundane level as evidenced by the findings of a researcher called Murphy. Roundworld would be difficult for them. They would be bewildered. Things would not work the way they expect. They are not flexible thinkers.'

'They were able to mess around with Mr Darwin's voyage!'

'By doing lots of small and rather stupid things at great effort, sir. They don't react well to adversity. They get petulant. From what Professor Rincewind tells us, many hundreds of them have to combine to perform even a simple physical action.'

He stood back and indicated some items laid out on a dining table.

'There is some evidence that Auditors, being embodiments of physical laws, find it hard to deal with nonsensical or contradictory instructions. Therefore, I have prepared these.'

He flourished something that looked like a table-tennis bat. On it were printed the words: 'Do Not Read This Sign.'

'That works, does it?' said the Dean, doubtfully.

'It's said to put their minds into a fugue state, Dean. They feel confused and alone, and evaporate instantly. Being alone means having a sense of self, and any Auditor that develops an individuality is said to die instantly.'

'And the catapult bows?' What are *they* for?' said the Archchancellor, slapping the Dean's hand off one of them.

'In addition, it is possible that a collective of Auditors with sufficient presence in the material world may develop crude physical senses, and so I have adapted some catapult bows to fire a mixture of intense, er, stimuli. Old references suggest chilli, essence of Wahoonie or Blissberry blossoms, but modern thinking inclines to Higgs & Meakins Luxury Assortment.'

'Chocolate?' said Ridcully.

'They don't like it, sir.'

'But those things can live in empty space and inside stars, man!'

'Where chocolate is significantly absent, sir,' said Ponder, patiently. 'They keep away from it. Also, it comes handily packed. They particularly don't like the Strawberry Whirl.'

Ridcully picked up a bow, pointed it at a wizard, and fired. There was a distant 'ow!'

'Hmm. Spreads nicely on impact,' he said. 'Well done, Mr Stibbons. I'm impressed. You are in charge.'

The Dean bridled at this. 'I protest! I am the Dean, when all is said and done!'

'Oh, all *right*, Dean, you can come! But, and I want to make myself *absolutely* clear, you are not to point anything at anything unless I give you a clear instruction, understood?'

'Yes, Mustrum,' said the Dean meekly.

'Furthermore, you will not, at any point, wave your weapon in the air and shout "choc and load". Is that clear? I say that because I can practically see the silly words forming in your head!'

'That's a vile calumny!' the Dean shouted.

'I hope so. Stibbons, wait here with the proctors and see no harm comes to Mr Darwin. Hex, you know where to send us. Invisibly, if you please!'

While Charles Darwin was sitting in a blue haze in Unseen University, a slightly younger Charles Darwin was staring out at the rain, noting idly that the rain sounded a little like whispering.

A drawback of invisibility is that no one can see you; it is in fact the main drawback if there is a group of you –

'– *that was my foot!*'

'*Who is that?*'

'*Look where you're going!*'

'*And what help will that be?*'

'*Keep it down, you fellows! He'll hear you!*'

At which point, the wall in the corner dissolved and brilliant light shone through. Beetles of all sizes and colours poured into the study in a shimmering torrent.

A figure that the wizards recognised stepped though the hole and looked around him with an air of amiable bewilderment. He had a slightly lopsided circlet of leaves on his head, and glowed with the light of deity.

'Mr Darwin?' he said, as the figure in the corner turned and stared. 'I understand you are studying evolution and are currently perplexed?'

'Look behind him!' Ridcully whispered.

The unseen wizards stared into the flickering hole. There was sand, and sea in the distance, a suggestion of moving shadows . . .

'After me,' Ridcully hissed, as an astonished Darwin dropped to his knees. 'Let's *get* them . . .'

The wizards poured through the camiloop, while behind them an elderly voice said: 'Of course, selection is, ahaha, anything but natural. Take, for example, a species of parasitic wasp . . .'

The sand boiled. Sometimes handfuls of it fountained into the air. One invisible person can move with stealth and speed. Half a dozen invisible people are an accident waiting to happen again and again.

'This is not being our finest hour,' said the voice of Ridcully. 'Every time I start to stand up someone else treads on me! Can't Hex sort this out?'

'We're back in the real world,' said the invisible Dean. 'Hex's power isn't so strong here. It'll take him some time to find us. Would you mind getting off my leg. Thank you so very much.'

'That's not me, I'm over here. I don't see *why* it's a problem. We were in another world, after all!'

'Roundworld is right inside the High Energy Magic Building,' said the Lecturer in Recent Runes. 'We're thousands of miles way, I suspect. Could I possibly suggest we all endeavour to crawl away in different directions? If you, Dean, head for that little bush with the red flowers, and Rincewind – where's Rincewind?'

'Here,' said a muffled voice from under the sand.

'Sorry . . . *you* head for that rock *there* . . .'

By degrees, with only the occasional curse, the wizards were able to get to their unseen feet.

'This is Mono Island, I recognise that mountain,' said Ridcully. 'Look out for—'

'Why didn't we just bop him on the head?' said the Dean. 'Just a tap on the noggin? Then we could have dragged him back here, end of problem.'

'But it's quantum,' said Rincewind. 'We have to deal with what's happened. If we stop it happening before it happens, the other things we've . . .' he hesitated. 'Look, it's quantum. Believe me, I'd prefer it the other way.'

'Anyway, you can't just bop gods on the head,' said Ridcully, now a faint outline against the distant ocean. 'It doesn't usually work and it causes talk. The other gods would be bound to hear about it, too.'

'So? None of them like him. They exiled him here after he invented the hermit elephant!' said the Dean, who was also fading into view.

'It's the look of the thing,' said Ridcully. 'They don't want to encourage deicide. Besides, look up there . . .'

'Oh dear,' said Rincewind. 'Auditors . . .'

A grey cloud was rolling down the mountain. As it neared, it contracted upon itself, growing darker.

'They've *learned* things,' said Ridcully. 'They've never done *that* before. Oh, well . . . Rincewind, first line of defence, if you please. And hurry!'

Rincewind, who'd always operated on the assumption that if you carried a weapon you were giving the enemy something extra to hit you with, held up a placard. It read: GO AWAY.

'Stibbons says it should work,' said Ridcully, uncertainly.

The Auditors drew closer, merging until, now, only half a dozen were left. They were dark, and full of menace.

'Ah, they probably aren't the reading sort, then,' said Ridcully. 'Gentlemen, it's chocolate time . . .'

It had to be said that the most of the wizards were not natural aimers. A spell went where you wanted it to go. You just had to wave in the general direction. They'd never learned to be serious about pointing.

Some shots went home. When several hit an Auditor it let out a thin scream and began to break up into its component robes, which then evaporated. But one, slightly large than the others, zigged and zagged through the tumbling chocolates. Auditors *did* learn here . . . and the wizards were running out of chocolate.

'Hold it,' said the Dean, pointing his bow.

The shape stopped.

'Ah,' said the Dean, happily. 'Ha, I expect you are wondering, eh, I expect you are wondering, indeed, if I have any chocolate left? And as a matter of fact I'm no—'

'No,' said the Auditor, drifting forward.

'What? Pardon?'

'I am not wondering if you have any chocolate left,' said the dark apparition. 'You have none left. The Higgs & Meakins Luxury Assortment comprises two each of: Walnut Whips, Strawberry Whirls, Caramel Bars, Violet Creams, Coffee Creams, Cherry Whips and Walnut Clusters and one each of Almond Delight, Vanilla Cup, Peach Cream, Coffee Fondue and Lemon Extravaganza.'

The Dean smiled the smile of a man whose Hogswatches had come all at once. He raised the bow.

'Then be so kind as to say good day to the Nougat Surprise!'

There was a twang. The sweet flew. For a moment the Auditor wavered, and the wizards held their breath. Then, with the slightest of whimpers, it faded into nothing.

'Everyone forgets the Nougat Surprise,' said the Dean, turning to the other wizards. 'I suppose it's because it's so irredeemably awful.'

There was nothing but the sound of the sea for a few seconds. Then:

'Er . . . well done, Dean,' said Ridcully.

'Thank you, Archchancellor.'

'A little too showy, nevertheless. I mean, you didn't have to *chat* to the thing.'

'I wasn't in fact sure if I *had* used the nougat,' said the Dean, still smiling. Quite an effort would be needed to wipe that smile away, Ridcully knew, and so he gave up.

'Good show, all the same,' he mumbled, and then raised his voice.

'If you can hear me, Hex . . . back to the Great Hall, please.'

Nothing happened. An important part of transferring matter *across* the world is the moving of an equivalent mass the other way. This can take a while.

Then an oak table, three chairs and two spoons crashed into the beach. A moment later, the wizards vanished.

TWENTY-TWO

FORGET
THE FACTS

 . . . IT'S THEORIES THAT MATTER.

Discworld does not have science as such. But it does have a variety of systems of causality, ranging from human intentions ('I'll just go out for a drink in the Mended Drum') to magical spells to a generalised narrativium that keeps local and general history close to the lines of 'story'. Roundworld does have science, but it's difficult to discover the extent to which it determines, modifies, affects people's actions – technology does, of course, but does science? Science does *affect* what we do, what we think, but it doesn't *change* what we do and think because so much of our basic knowledge is simply accepted scientific 'fact'.

Well, actually not 'fact', but theory.

We search for theories because they organise facts. We do this, according to *The Science of Discworld II*, because we are really *Pan narrans*, the storytelling ape, not *Homo sapiens*, the wise man. We invent our own stories to help ourselves to live. For this reason we are not reliable when we collect 'facts' for scientific purposes. Even the best scientists, and certainly the paid help and the student employees, are so full of what they *want* to find that there's no way that what they do find can relate to the real world more than to their own prejudices, biases, and wishes. However, we were all told at school that 'science's facts are reliable', but that its theories – and even more so its working hypotheses – are and were constantly subject to criticism, and therefore to change. It was explained to us that Newton had been supplanted by Einstein, Lamarck by Darwin, Freud by Skinner . . . So we were told that theories were constantly being supplanted, but that the observations on which they were based were reliable.

This is the reverse of the truth.

No teacher pointed out that many, perhaps most, of the basic assumptions of our intellectual world were scientific theories that had *survived* criticism . . . from the place of Earth and Sol in the Milky Way galaxy to the fertilisation theory of human conception to subatomic physics producing atom bombs . . . to Ohm's Law and the electrical energy grid, to medical tricks like the germ theory of disease, all the way to X-rays and MRI (magnetic resonance imaging), not to mention chemical theories that reliably gave us nylon, polythene and detergents. These theories go unnoticed because they have become defaults, so completely accepted as 'true' that we fail to paint them with emotional tags, and simply build them into our intellectual toolkit. Even though no teacher pointed out that they were scientific successes, they constitute much of the (regrettably but unavoidably) uninspiring parts of school science.

On these foundation beliefs we hang such glittering flesh as visits to Mars, new fertility techniques like ICSI, fusion power, new bactericides for kitchen surfaces – and for a minority of the more imaginative children, the wide and wonderful worlds of science fiction.

The theories of science, then, particularly the totally accepted ones like sperm-egg conception, polythene, and Earth-orbits-Sun, are good reliable science. They are continually tested against the real world when babies are conceived in fertility clinics, when people do the washing-up, or when astronauts circle the Earth in sunlight and shadow. An enormous mass of Roundworld science is built into our everyday world, and it's mostly reliable.

But there is also a whole mass of science that is incomprehensible to nearly everybody, which pretends that it's The Answer for all kinds of technical or philosophical issues, and which supports experts. Quantum theory is the classic case, relativity is a touch more accessible, but subatomic physics and most of medicine, aeronautics and automobile engineering, soil chemistry and biology, statistics, and the higher reaches of economics, are all subjects that nearly everybody is content to leave to the experts. Mathematics has a strange position, similar but with its own peculiar stance akin to revealed religion – mostly because it has been presented from school onwards as an arcane craft whose practitioners are the only humans with access to Platonic truths.

Then there are the quasi*-sciences like astrology, homeopathy, reflexology, and iridology, which simply can't work. They should be sharply distinguished from odd, often ancient practices like acupuncture, osteopathy and herbal treatments, which work sufficiently often but have a theoretical base that is poorly worked out in scientific terms. Many people are attracted by their homespun mix of myth and mysticism (which are all the more impressive because the treatment sometimes *works*), and feel that a modern scientific investigation would somehow spoil them. It would certainly poke some holes in the traditional rationalisations, but in all likelihood it would make the treatments even better. Whereas the quasi-sciences would be (indeed, already have been, not that everyone's noticed) demolished.

To end that list, we add evolutionary biology, a very well-established set of models founded in the fossil record, chromosomes, and DNA, which explains similarities and differences among today's living creatures much more elegantly and effectively than its creationist or intelligent-design rivals. Nevertheless, a very large proportion of people – especially Christians in the American Mid-West, Muslims in fanatically Islamic cultures, and fundamentalist believers in general

---

* Pronounced 'crazy'.

– deny that humans evolved. To them, their own brand of authority trumps the scientific evidence, or their 'common sense' renders the whole concept laughable. 'I ain't kin to no ape!' was the explanation given by a young schoolgirl at one of Jack's *Life on Other Planets* lectures, when the teacher asked her why she didn't believe in evolution.

There is a general human propensity, of which much use is made in the Discworld books, to set up accepted, unexamined mental backgrounds. Mostly these result from the Make-a-Human-Being kits that each human culture inflicts on its members as they grow up through childhood and adolescence. Each of us is the result of a learning process, only a tiny fraction of which is overt 'education' by professional teachers. The kit includes nursery rhymes, songs, stories, the personification of nursery animals (sly foxes, wise owls, industrious litter-collecting Wombles) and human roles from fabulous postman and princess up to crime-fighting Batman and Superman. All these have their place in the unexamined basis of our day-to-day thoughts and actions. A possible explanation for Princess Diana's undeniable popularity with the British public – indeed the world – is that she, unlike 'real' royalty, had imbibed the *popular* impression of What Princesses Do as distinct from the authentically royal version. So she did what *we* had all learned that real princesses do, she looked and behaved like an icon, not like genuine royalty.

Sophisticated human beings, citizens like us – and indeed like tribesmen and barbarians* in today's world, nearly all of whom have heard of Superman, Tarzan, Ronald MacDonald – all have this hotchpotch

* This is a special usage devised by the anthropologist Lloyd Morgan in the 1880s, picked up by John Campbell Jr in an *Analog* editorial in the 1960s, then by Jack in *The Privileged Ape*: for tribal humans, everything is traditional, mandatory or forbidden; for barbarians, action is driven by honour, bravery, modesty, defiance of precedent; for citizens, some roles are tribal, some barbarian, we choose.

of images, models, phobias, inspirations and villains. Our day-to-day experience gives us a self whose memory-train is a succession of scenes, thoughts, experiences, and passions, all painted à la Damasio with emotional tags that say 'Great!', 'Do This Again When I Can!' or 'Avoid At All Costs!' when we recall them. But these sit upon a great mass of mostly unexamined structural human material, that labels us as Western Twentieth-Century Biologist or Ghetto Rabbi or Roman Centurion or Seventeenth-Century French Courtesan, or, for most people most of the time, Exploited Peasant.

Each of those roles has a different set of emotional labels for money, for priests, for sex, for nakedness, for death, and for birth. Most people, until quite recently, underpinned that unexamined set of beliefs with a theist (personal, humanlike) God or gods, or a deist (Something Up There with Extraordinary Powers) god-structure, so the emotional tags on important memories have been strongly God-flavoured. When we remember them they may be sins, atonements, redemptions or trials. They may be mitzvahs (blessings) or revenges or charities. Religions, in bringing us into our cultures *via* their Make-a-Christian or Make-a-Maya kit, put different-strength labels on, for example, human sacrifice, so that it has a whole host of associations in adult life. Our adult prejudices, and our scientific theories, go in on top of this crazy mishmash of historical errors, badly understood schooling, mathematics and statistics that barely make sense to us, God-stories of causality and ethics, and educational lies-to-children that permit the teacher to disengage his brain in response to children's questions.

This mental mishmash is well illustrated by our changing attitudes to Mars. Mars was known to the ancients as a 'wandering star', a planet; its reddish colour had bloody associations, so the Romans associated it with their god of war. It also acquired a connection with war in astrology, where the visible stars and planets all had to *mean* something. We're going to look at a lot of different associations with Mars,* as myth and rationality engaged with the red planet, as stories by the hundred employed Mars and Martians, and as the scientific picture of Mars changed over the centuries.

We shouldn't ask 'which is the true Mars?' We become larger humans by considering all of these aspects; from that stance there really isn't a true, real, objective planet for our minds to engage with usefully. Our simple, thin causal lines can't comprehend a real astronomical object, even a world which is actually out there so that we can see it. The 'it' we see can be the disc whose apparent lines Giovanni Schiaparelli called 'canali', which excited Percival Lowell (whose grasp of Italian seems to have been slight, since the word means 'channels') to see them as engineered canals. He wrote *Mars as the Abode of Life*, and this laid the foundation for the folk Mars of the twentieth century.

Between the World Wars, everybody in the West, and many in the East, looked into the night sky and saw inimical Martians, a mental residue of that 1920s picture of a drying, dying Mars. The image was overlaid by the *War of the Worlds* picture of envious, grim, disgusting tripod Martians invading Earth (or at least England). There was a more romantic overlay for many of those out camping, or sleeping out under the stars: *Barsoom*. Edgar Rice Burroughs, familiar because of his Tarzan stories, invented a Mars whose dried-out seabeds were home to green Martian warrior hordes, six-legged centaur-like creatures whose egg-incubators were visited regularly. John Carter, an American ex-confederate army officer, had wished himself on to Mars, been captured by the green warriors but soon found himself married to a red Martian princess.† Stanley Weinbaum's *A Martian Odyssey* added more dimensions: the Martian called Tweel,

---

* Not quite including the confectionery, which was the surname of the originator; he came to England from the USA, and invented M&Ms too. That stands for 'Mars and Mars'.
† Also egg-laying. Jack, reading Burroughs when young, was disturbed by the idea of their marriage bed . . .

who made long hops and landed on his nose, the hypnotic predator that showed you your most desirable images, and attempts at a
gosh-wow desert ecology. Then there were stories of Martians coming to Earth, pretending to be human . . . and humans attempting to
interact with a more or less mystical ancient Martian civilisation.

The best known, perhaps the best crafted of these romantic-
mystical portrayals of crude, lumbering Earthmen, insensitive to the
ethereal beauties of the Martian crystal cities, were Ray Bradbury's.
In the 1950s and 1960s his tales were read by many outside the
fantasy/SF world, and they appeared in widely read magazines like
Argosy as well as in SF pulps in railway station bookstores. They laid
the mystical ancient Martian foundation for Robert Heinlein to build
the most potent of all these Martian tales, *Stranger in a Strange Land*.
Michael Valentine Smith had been a foundling on Mars, brought up
and trained in their culture by the ancient Martians. He came to Earth,
founded a commune of friends – 'Water Brothers' – and started a religion whose 'grokking the fullness' of everyday events, from sex to
science to swimming, spread to communities of readers. There was
a tragic, well-publicised association with the murderous Manson
killers, who had used this book as their mantra, but this didn't harm
sales, and the ancient mystical Martians became the standard image.

Then we learned that Mars has no atmosphere to speak of, that it
is cold, dry, laden with frozen carbon dioxide, to the extent that the
'icecaps' were probably dry ice. Our machines visited Mars, looked
for 'life', and found strange chemistry because we inevitably asked
the wrong questions. The 'canals' died in the public mind, replaced
by craters and gigantic volcanoes.

We have now visited again, and it seems that ancient, wet Mars
may have been a reality, there may be at least bacterial life forms
under the sand . . . Much is not yet clear, but what is clear is that our
image of Mars has changed yet again.

Each of us has a variety of associations with Mars. When we weave
these many different interpretations and imaginations together, we

become a different, wiser kind of creature. As for all of our different Marses . . . well, those are toys of our imaginations, as we grok the red planet's fullness.

If Mars seems a bit of a digression, consider those twin icons of evolution, the archaeopteryx and the dodo. In folk-evolutionary thinking, the archaeopteryx is the ancestor of all the birds, and the dodo is the bird that went extinct about 400 years ago. 'As dead as a dodo.' Again, our thinking about these iconic creatures is heavily daubed with unchallenged assumptions, myths, and fictional associations.

We mentioned archaeopteryx in Chapter 36 ('Running from Dinosaurs') of *The Science of Discworld*, second edition. We think of it as the ancestral bird because it is a dinosaur-like animal with bird-like feathers . . . and *it was the first one to be found*. However, by the time of archaeopteryx there were plenty of genuine birds around, among them the diving bird *Ichthyornis*. Poor old archaeopteryx arrived on the scene far too late to be 'the' bird ancestor.

The recent amazing 'dinobird' discoveries in China – transitional creatures part way between dinosaurs and birds – have totally changed scientists' view of bird evolution. At some stage some dinosaurs started to develop feathers, though they couldn't then fly. The feathers had some other function, probably keeping the animal warm. Later, they turned out to be useful in ·wings. Some dinobirds effectively had four wings – two at the front, two at the back. It took a while before the standard 'bird' body-plan settled down.

As for the dodo – we all know what it looked like, right? Fat little thing with a big hooked beak . . . Such a famously extinct creature must be well documented in the scientific literature.

No, it's not. What we have is about ten paintings and half a stuffed specimen.* We have more specimens of the archaeopteryx than we

---

* Rajith Dissanayake, 'What did the Dodo look like?' *Biologist* 51 (2004), 165–8.

do of the dodo. Why? The dodo went *extinct*, remember? And it did so before science really got interested in it. So few people recorded it, or studied it. It was *there*, not requiring special attention, and then it wasn't, and it was too late to start studying it. It isn't even certain what colour it was – many books say 'grey', but it was more likely brown.

Yet, we all know exactly what it looked like. How come? Because we all saw it illustrated by Sir John Tenniel in *Alice's Adventures in Wonderland*.

Say no more.

The great strength of Discworld narrative is that it makes fun of just those places where 'education' has left us feeling a bit vulnerable: where we change the subject in the pub, or when our five-year-old asks us those probing questions. A running joke throughout the *Science of Discworld* series is what grammarians call 'privatives'. These are concepts that our minds seem quite happy with, even though a moment's thought shows that they're complete nonsense. Chapter 22 of *The Science of Discworld* discussed this notion, and we recap briefly.

It is entirely normal to speak of 'cold coming in the window' or 'ignorance spreading among the masses'. The *opposites* of these concepts, heat and knowledge, are real, but we've dignified their absence with words that do not correspond to actual things. In Discworld, we find 'knurd', which is super-sober, as far from ordinary sober as drunk is in the alcoholic direction. There are jokes about the speed of dark, which must be faster than the speed of light because dark has to get out of the way. On Discworld, Death exists as a (perhaps *the*) major character, but on Roundworld that word refers only to the absence of life.

People habitually label the absence of something with a word, instead of (or as well as) its presence: such words are the afore-mentioned privatives.

Sometimes this habit leads to mistakes. The classic case was the label 'phlogiston', the substance that appears to be emitted by burning materials. You can *see* it coming out as smoke, flame, foam . . . It took many years to demonstrate that burning was an intake of oxygen, not the emission of phlogiston. During the intervening period, many people had demonstrated that when metals burned they got heavier, and had therefore argued that phlogiston had negative weight. These were clever people; they weren't being stupid. The phlogiston idea really did work – until oxygen supplanted its explanations, and alchemists suddenly found that the paths into rational chemistry were easier.

Privatives are often very tempting. In *What is Life?*, a short book published in 1944, the great physicist Erwin Schrödinger asked precisely that question. At that time the Second Law of Thermodynamics – everything runs down, disorder always increases – was thought to be a fundamental principle about the universe. It implied that eventually everything would become a grey, cool soup of maximum entropy, maximum disorder: a 'heat death' in which nothing interesting could happen. So in order to explain how, in such a universe, life could occur, Schrödinger claimed that life could only put off its individual tiny heat death by imbibing negative entropy, or 'negentropy'. Many physicists still believe this: that life is unnatural, selfishly causing entropy to increase more in its vicinity than it would otherwise do, by eating negentropy.

This tendency to deny what is happening before our very eyes is part of what it is to be human. Discworld exploits it for humorous and serious purposes. By building Discworld flat, Terry pokes fun at flat-Earthers; rather, he recruits his readers into a 'we all know the Earth is round, don't we?' fellowship. The Omnians' belief in a round Disc, in *Small Gods*, adds a further twist.

We want to put what rational people are coming to believe into a general human context, so let's look at what everyone believes. In these days of fundamentalist terrorists we would do well to understand why a few people hold beliefs that are so different from the rational. These unexamined beliefs may be vitally important, because the ignorant people who espouse them think that they provide a good reason for killing us and our loved ones, even though they have never considered alternatives. People Who Know The Truth, by heredity or personal revelation or authority, are not concerned with logic or the validity of premises.

Nearly everybody who has ever lived has been one of those.

There have been a few sparse times and places – and we are hoping that the twenty-first century will host a few of them – in which onlookers are more ready to believe a disputant who is unsure, than one who is certain. But in today's politics, changing your mind in response to new evidence is seen as a weakness. When he was Vice-Chancellor at Warwick University, the biologist Sir Brian Follett remarked: 'I don't like scientists on my committees. You don't know where they'll stand on any issue. Give them some more data, and they change their minds!' He understood the joke: most politicians wouldn't even realise it *was* a joke.

In order to discuss the kinds of explanation and understanding that are going to have future values, we need at least a simple geography of where human beings pin their faiths now. What kinds of world picture are most common? They include those of the authoritarian theist, the more-or-less imaginative theist, the more critical deist, and various kinds of atheists – from Buddhists and the followers of Spinoza to those, including many scientists and historians, who simply believe that the age of religion is behind us.

Most human beings of the last few millennia seem to have been authoritarian theists, and we still have many of them in our world; perhaps they are still a majority. Does this mean that we must give intellectual 'equal time' to these views (plural, because they're all very different: Zeus, Odin, Jahweh . . .), or can we just dismiss them all with 'I have no need of that hypothesis', as Laplace supposedly said to Napoleon. Voltaire, aware that God making man in His image meant that God's nature might be deduced from man's, thought it at least possible that God has mischievously misinformed us about reward and punishment. Perhaps sinners are rewarded by Heaven and saints are given a taste of Hell. Our view is that all the various authoritarian theists are the contemporary bearers of an extremely successful memeplex, a package of beliefs designed and selected through the generations to ensure its own propagation.

A typical memeplex is the Jewish *shema*: 'And these words . . . you shall teach them to your children, muse on them when you get up and when you lie down . . . Write them upon the doorposts of your house and upon your gates.' Like e-mail chain letters that threaten you with punishment if you fail to send them on to many friends, and with 'luck' if you do send them on, the world's great religions have all promised pleasures for committed believers and transmitters, but pain for those who fail to adhere to the faith. Heretics, and those who leave the faith, are often killed by the faithful.

We can easily understand how such beliefs, bolstered as they are from within, have been retained throughout the generations. The promise of an afterlife, espoused by all the sensible people around you, makes many of this life's sorrows easier to bear. And, as we've seen in recent years, belief in a Paradise for those who die fighting for the faith in a Holy War makes you pretty well invincible.* Such invincibility is a side-effect of the memeplex's belief tactics, not a certification of the truth of the bomber's faith. Especially given that nearly all of those who share the bomber's faith (Islam, Catholicism . . .) deny that their beliefs justify killing unbelievers.

This plurality of theist beliefs, especially in today's mixed-up world with its different cultures and multicultures, encourages a more critical belief in authority, and usually a willingness to admit

---

* Although it does seem a little strange that Palestinian terrorists protect their genitals, for use in Paradise, when setting themselves up as suicide bombers.

commonality with other theists. Such common ground encourages the integration of different cultures. Many minorities are assimilated and disappear, but others react by emphasising their individuality. Some of the latter, like the Thuggee worshippers of the death goddess Kali in nineteenth-century India, and the recent al-Qaeda terrorists, gain temporary notoriety that seems to be a triumph of their faith. However, this is usually self-defeating in the longer term. In any event, the number of deaths is no comment, plus or minus, upon the truth of the beliefs that these thugs hold. The faith of these militant minorities sometimes gains sharpness and even subtlety, but it is usually subordinated to the day-to-day exigencies of the violent lives they lead.

Many great scientists, for example Galileo, were ridiculed when they proposed new insights into the natural world. Scientific crackpots often deduce that because their work is being ridiculed, they must be the new Galileo, but that doesn't follow. Similarly, men of violence often try to validate their 'martyrdom' by comparing themselves to ancient Christians or ghetto Jews, and again the logic is flawed. There is no rational reason to accept any of their gods as part of the real universe, however helpful that belief might be to some people as regards day-to-day living. Despite that, many clever, honest people do feel that a God is necessary to their understanding of how everything is organised. Once a memeplex has caught you, it's hard to escape.

We have a little more sympathy with deists, who mostly seem to believe that the universe is extraordinarily complex, yet possesses an overall simplicity, and that this points to some celestial guardian who looks after the whole thing and gives it meaning. Ponder Stibbons and Mustrum Ridcully, in their different ways, edge towards deism; they want to feel that 'someone' is at the helm. If challenged, deists usually deny the anthropomorphic character of this guardian, but they still retain a belief in the ability of individual people – perhaps individual 'souls' – to relate directly with whoever or whatever is in charge. We personally doubt that such apparent interactions, whether attained by meditation or prayer, are more than self-delusion. But we are happy to live on the same planet as people who believe that they are in direct contact with ultimate causality, however unscientific we may feel that belief to be.

There is a growing minority of thoughtful people who have given up on the idea of a personal, anthropomorphic God altogether. Some, particularly among Buddhists and Taoists, retain the mystical/metaphysical stance that is characteristic of religion, and consider the 'scientific' world to be subservient to a mystical true picture more closely related to subjective experience. In contrast, those of us who were persuaded by Spinoza's rejection of an anthropomorphic God, not least because neither the universe nor an omnipotent deity can exist without being coextensive with everything there is, see the scientific view as exposing both the nature of god, if that is our belief, through the laws by which things work, and the workings of the universe itself.

Many scientists, particularly those whose endeavours relate closely to the real world, like geologists, astronomers, biologists, ecologists, and polymer chemists, avoid the mystical approach and see their own speciality as exemplifying a complex slice of the universe, with many emergent properties that are not predictable from the detailed substructure. Other scientists, particularly those devoted to reductionist explanations, like physicists, astrophysicists, physical chemists, molecular biologists, and geneticists, retain a version of the mystical approach, but try to explain higher level behaviour in terms of the substructure. Tellingly, many scientists who work at the 'coal face' of the subject generally have a respect for the unknown possibilities that the universe might throw at them, whereas workers in more abstract realms like quantum theory have a tendency to go all mystical about their own understanding, or lack of it.

Most human attempts at an explanation try to find a thin causal chain of logic and narrative, leading from things we accept to whatever it is we are trying to explain. This type of story appeals to human minds, but it is usually an oversimplification, and it leads to serious misunderstandings. The typical television science programme, where a single individual is held to be responsible for some big 'breakthrough', paints a wildly inaccurate picture of the incremental process by which most scientific advances are made. Unicausal explanations make nice stories, but fail to capture the complexities of the real world. The most effective explanations are often very varied, and it's a good idea to find a lot of different ones, if they're available. Physicists searching for a unification of relativity and quantum theory should perhaps bear in mind the possibility that any unification may turn out to be less effective than the two separate theories, each safely confined to its own domain. Only if you can get several theories to compete, in your mental territory, can you begin to distil understanding.

TWENTY-THREE
THE GOD OF EVOLUTION

'DOING WELL BUT LOTS STILL TO BE DONE!' barked Ridcully, striding out of the magic circle into the Great Hall. 'Everything all right, Mr Stibbons?'

'Yes, sir. You didn't try to stop the God of Evolution talking to Darwin, did you?'

'No, you said we shouldn't,' said Ridcully briskly.

'Good. It had to happen,' said Ponder. 'So all we need to do now is persuade Mr Darwin—'

'I've been thinking about that, Stibbons,' Ridcully interrupted, 'and I have decided that you will now take Mr Darwin to meet our God of Evolution on his island,' said the Archchancellor. 'It's quite safe.'

Ponder went pale. 'I'd really rather not go there, sir!'

'However, you will, because I am Archchancellor and you are not' said Ridcully. 'Let's see what he thinks of the wheeled elephant, eh?'

Ponder glanced at Darwin, still in the blue glow of stasis. 'That's very dangerous, sir. Think of what he'll be seeing! And it would be quite unethical to remove the memories that—'

'I know I *am* Archchancellor, it's written on my door!' said Ridcully. 'Show him his god, Mr Stibbons, and leave the worrying to me. Quickly, man. I want this all wrapped up by dinnertime!'

A moment after Ponder and Darwin left, a small boulder and quite a lot of sand appeared and slid across the tiles of the Great Hall.

'Well done, Mr Hex,' said Ridcully.

+++ Thank you, Archchancellor +++ Hex wrote.

'I was kind of hoping we'd get the chairs back, though.'

+++ I will see what I can do next time, Archchancellor +++

And on Mono Island, Charles Darwin picked himself up from the beach and stared around.

'Does this lend itself to any rational explanation, or is it more madness?' he said to Ponder. 'I have cut my hand quite badly!'

At which point two little leaves pushed themselves out of the ground near his foot and, with amazing speed, became a plant. It threw up more leaves, then developed a single red flower which opened like an explosion and died like a spark to produce one single seed, which was white and fluffy.

'Oh, a bandage plant,' said Ponder, picking it. 'Here you are, sir.'

'How—' Darwin began.

'It just understood what you needed, sir,' said Ponder, leading the way. 'This is Mono Island, the home of the God of Evolution.'

'A *god* of evolution?' said Darwin, stumbling after him. 'But evolution is a process inherent in—'

'Ye yes, I know what you're thinking, sir. But things are different here. There *is* a god of evolution and he . . . *improves* things. That's why we think everything here is desperate to get off the island, poor creatures. Somehow they know what you want and evolve as fast as they can in the hope you'll pick them to take away.'

'That is not possible! Evolution needs many thousands of years to—'

'Pencil,' said Ponder, calmly. A tree nearby shivered.

'Actually, the pencil bush breeds true in the right soil,' Ponder went on, walking over to it. 'We've got some of these at the University. And the Chair of Indefinite Studies kept a cigarette tree going for months, but they got very tarry. Once most of them get far enough away from Mono Island they stop trying.' He held one out. 'Would you like a ripe pencil? They're quite useful.'

Darwin took the slim cylinder Ponder had plucked from the tree. It was warm, and still slightly moist.

'This is Mono Island, you see,' said Ponder, and pointed to the small mountain at the far end of the island. 'Up there is where the god lives. Not a bad old boy, as gods go, but he will keep *changing* things all the time. When we met him he—'

The bushes rustled, and Ponder dragged the bemused Darwin aside as something rattled down the path.

'That's a giant tortoise!' said Darwin, as it trundled past. 'That at least is something – oh!'

'Yes.'

'It's on wheels!'

'Oh, yes. He's very keen on wheels. He thinks wheels should work.'

The tortoise turned quite professionally and rolled to a halt by a cactus, which it proceeded to eat, daintily, until there was a hiss and it sagged sideways.

'Oh,' said a voice from the air. 'Bad luck. Tyre bladder punctured. It's the everlasting problem of the strength of the integument versus the usage rate of the mucus.'

A skinny, rather preoccupied man, dressed in a grubby toga, appeared between the two of them. Beetles orbited him like wonderful little asteroids.

'Deposition of metal may be our friend here,' he said, and turning to Ponder as if to another old friend he went on: 'What do you think?'

'Ah, um, er . . . do you really need all that shell?' said Ponder, hurriedly. Beetles, bright as tiny galaxies, landed on his robe.

'I know what you mean,' said the old man. 'Too heavy, perhaps? Oh . . . you seem familiar, young man. Have we met before?'

'Ponder Stibbons, sir. I was here a few years ago. With some wizards,' said Ponder, with care. He'd quite admired the God of Evolution, until he'd found that the god considered the cockroach to be the peak of the evolutionary pyramid.

'Ah, yes. You had to leave in such a hurry, I recall,' said the god, sadly. 'It was—'

'You! . . . you appeared in my room!' said Darwin, who'd been staring at the god with his mouth open. 'There were beetles everywhere!' He stopped, his mouth opening and shutting for a while. 'But you certainly are not . . . I thought you—'

Ponder was ready for this.

'You know about Olympus, sir?' he said quickly.

'What? This is *Greece?*' said Darwin.

'No, sir, but we've got lots of gods here. This, er, gentleman isn't, as you might put it, *the* god. He's just *a* god.'

'Is there a problem?' said the God of Evolution, giving them a worried smile.

'*A* god?' Darwin demanded.

'One of the nice ones,' said Ponder quickly.

'I like to think so,' said the god, beaming happily. 'Look, I need to check on how the whales are doing. Why don't you come up the mountain for tea? I love to have visitors.'

He vanished.

'But the Greek gods were myths!' Darwin burst out, staring at the suddenly empty space.

'I wouldn't know about that, sir,' said Ponder. 'Ours aren't. On this world, gods are extremely real.'

'He came through the wall!' said Darwin, pointing angrily at the empty air. 'He told me that he was immanent in all things!'

'He tinkers a lot, certainly,' said Ponder. 'But only here.'

'*Tinkers!*'

'Shall we take a little walk up Mount Impossible?' said Ponder.

Most of Mount Impossible was hollow. You need a lot of space when you are trying to devise a dirigible whale.

'It really should work,' said the God of Evolution, over tea. 'Without that heavy blubber and with an inflatable skeleton of which, I must say I am rather proud, it should do well on the routes of migratory birds. Larger maw, of course. Note the cloud-like camouflage, obviously required. Lifting is produced via bacteria in the gut which produce elevating gases. The dorsal sail and the flattened tail give a reasonable degree of steerability. All in all, a good piece of work. My main problem is devising a predator. The sea-air ballistic shark has proved quite unsatisfactory. I don't know if you might have any suggestions, Mr Darwin?'

Ponder looked at Darwin. The poor man, his face grey, was staring up at the two whales who were cruising gently near the roof of the cave.

'I beg your pardon?' he said.

'The god would like to know what could attack this,' Ponder prompted.

'Yes, the grey people said you were very interested in evolution,' said the god.

'The grey people?' said Ponder.

'Oh yes, *you* know. You see them flying around sometimes. They said someone really wanted to listen to my views. I was so pleased. Lots of people just laugh.'

Darwin looked around the celestial workshop and said: 'I cannot see anything to laugh at in an elephant with sails, sir!'

'Exactly! It was the big ears that gave me the clue there,' said the god cheerfully. 'Making them bigger was simplicity itself. It can do twenty-five miles an hour across the open veldt in a good wind!'

'Until a wheel bursts,' said Darwin, flatly.

'I'm sure once they get the idea it will all work,' said the god.

'You don't think it might be better to let things evolve by themselves?' said Darwin.

'My dear sir, it's so *dull*,' said the god. 'Four legs, two eyes one mouth . . . so few are prepared to *experiment*.'

Once again Darwin looked around the glowing interior of Mount

Impossible. Ponder watched him take in the details: the cage of web-winged octo-monkeys that in theory could skim across the canopy for hundreds of yards, the *Phaseolus coccineus giganticus* that actually bred true, if there was any possible use for a beanstalk that could grow half a mile high . . . and everywhere the animals, often in stages of assembly or disassembly but all quite contentedly alive in a little mist of holiness.

'Mr, er, Stibbons, I should like to go . . . home now, please,' said Darwin, who had gone pale. 'This has all been most . . . instructive, but I should like to go home.'

'Oh dear, people are always rushing off,' said the god, sadly. 'But still, I hope I have been of help, Mr Darwin?'

'Indeed, I believe you have,' said Darwin, grimly.

The god accompanied them to the mouth of the cave, beetles streaming behind him in a cloud.

'Do call again,' he said, as they wandered off down the track. 'I do like to—'

He was interrupted by a noise like all the party balloons in the world being let down at once. It was long and drawn out and full of melancholy.

'Oh no,' said the God of Evolution, hurrying back inside, 'not the whales!'

Darwin was silent as they walked to the beach. He was even more silent as they passed the wheeled tortoise, which was limping in circles. The silence was deafening when Ponder summoned Hex. When they appeared in the Great Hall his silence, apart from a brief scream during the actual travelling, was a huge infectious silence that was contagious.

The assembled wizards shuffled their feet. Dark rage radiated off their visitor.

'How did it go, Stibbons?' whispered Ridcully.

'Er, the God of Evolution was his usual self, sir.'

'Was he? Ah, good—'

'I wish, very clearly, to awaken from this nightmare,' said Darwin, abruptly.

The wizards stared at the man, who was quivering with rage.

'Very well, sir,' Ridcully said quietly. 'We can help you wake up. Excuse us a moment.'

He waved a hand; once again the blue shimmer surrounded their visitor. 'Gentlemen, if you please?'

He beckoned to the other senior wizards, who clustered around him.

'We *can* put him back without him having any memory of anything that happened here, right?' he said. 'Mr Stibbons?'

'Yes, sir. Hex could do it. But as I said, sir, it wouldn't be very ethical to mess around with his mind.'

'Well, I wouldn't like anyone to think we're unethical,' said Ridcully firmly. He glared around. 'Anyone object? Good. You see, I've been taking to Hex. I'd like to give him something to remember. We owe him that, at least.'

'Really, sir?' said Ponder. 'Won't it make things worse?'

'I'd like him to know *why* we did all this, even if it's only for a moment!'

'Are you sure that's a good idea, Mustrum?' said the Lecturer in Recent Runes.

The Archchancellor hesitated. 'No,' he said. 'But it's mine. And we're going to do it.'

TWENTY-FOUR

A LACK OF SERGEANTS

 WHAT WAS IT ABOUT VICTORIAN ENGLAND, and what led up to it, that made it so progressive, inventive and innovative? Why was it so different from Russia, China, and all the other nations that seem to have stagnated during the nineteenth century – accumulating wealth, but lacking a middle class full of engineers, sea captains, clerics, and scientists? We would not expect there to be one simple answer, one trick that Victorian England discovered but other nations did not. That would satisfy the innate human wish for a single thin causal chain, but as we've seen, history doesn't work like that.

Equally, though, it's unsatisfying just to list lots of possible contributory causes – the East India Company . . . Harrison's excellent chronometer, which helped to make the British Empire so profitable and made aristocratic families send their younger sons fairly safely out into the Empire, from which they came back wiser and richer . . . Quakers and other nonconformist sects, which were tolerated by the Anglican Church . . . the Lunar Society's progeny, including the Royal Society and the Linnaean Society . . . the College of Apprentices . . . Parliament and the pretence of democracy, so that a middle class could rise from the merging of junior aristocrats who came back from the Empire to found pickle factories in Manchester . . . artisans who were coming into towns looking for satisfying jobs.

311

We could make the list ten times longer, though in most cases we wouldn't be sure about genuine causal connections. And even with ten times as many 'causes', we would still have to say 'all of the above'.

Are such factors a cause of historical differences, or a consequence? That's not a sensible question if you insist on a yes/no answer – very probably the answer should be 'both'. A modern analogue would be to ask whether today's space-oriented engineers and scientists are a cause of the success of space films and nailed-down science-fiction stories – or did the early scientifically oriented SF stories, with their sense of wonder at the sheer vastness and mystery of outer space, fire those engineers, when young, with the desire to turn fiction into fact? It must have been both, of course.

The early Victorian apprentices in pottery, ironworking, brick firing, and even bricklaying were respected by, and respected, their masters. Together they laid down enduring monuments for future generations. Similarly, early trains and canals connected all the major cities, and connected factories to their suppliers and customers. This transport system paved the way to the wonderful economic network that Edwardian Britain inherited from the Victorians. These systems were not static, to be admired for what they had achieved. They were dynamic, they changed, they were processes as much as achievements. They changed the way succeeding generations thought about where and how they lived. Even today, our cities rely heavily on what the Victorians built, especially when it comes to sewerage and water supplies.

The resulting changes in thinking fuelled further changes. The combination of cause and consequence is an example of what we have elsewhere called *complicity*.* This phenomenon arises when two conceptually distinct systems interact recursively, each repeatedly changing the other, so that they co-evolve. A typical outcome is that

* See Jack Cohen and Ian Stewart, *The Collapse of Chaos* (Viking, 1994).

together they work their way into territory that would have been inaccessible to either alone. Complicity is not mere 'interaction', where the systems join forces to achieve some joint outcome, but are not themselves greatly affected as a result. It is far more drastic, and it changes everything. It can even erase its own origins, so that neither of the original separate systems remains.

The social innovations that were (arguably but not solely) triggered by Victorian ingenuity and drive are just like that. Because there was selection, and because the best growth often occurs in the best run and best designed parts of growing systems, there was recursion. The next generation was inspired by the previous generation's successes, *and* their noble mistakes, and built a better world. What we might call the Channel Tunnel Syndrome occurs quite often in capitalist, democratic societies, but not in totalitarian states or even in nations like, say, today's Arab states or twentieth-century India. And particularly not in nineteenth-century Russia or China: both were rich, but they had no respectable middle class.

The Victorian middle class was respected both by the workers whose lives they exploited – and opened up – and by the aristocrats, whose increasingly international outlook was progressively integrated with trade. Russia and China had political systems without an economically powerful, shareholding middle class, which could start or follow fashions, and support romantic, visionary ventures. Today, the British will still support a Channel Tunnel venture or a *Beagle-2* Mars lander, because such things are romantic and possibly heroic, even though they are unlikely to be very profitable. A lengthy historical record shows very clearly that the first attempt at any major tunnel usually collapses financially – though after the tunnel is successfully built – often after a long series of attempts to shore up a failing enterprise. Then the ruins are bought for a song, occasionally nationalised or considerably financed by government or some other major capital source, and the resulting business can stand on the shoulders of the first. Only some rather strained economics has so far kept the original companies involved in the Channel Tunnel in business, at least on the British side of the Channel where everything was done by private enterprise.

Some projects are so romantic, so attractive in concept but so very difficult in execution, that three or four attempts are needed for them to acquire momentum. It is recursive structure of the complicit kind that keeps them afloat.* Telford's bridges are famous, as are so many of his other engineering works; his ability to capitalise on his successes was the result, and the cause, of his fame, which was achieved by what would now be called 'networking' among aristocrats, government ministers, and pickle manufacturers. He was, as they said, famous for being famous. In America similar enterprises were measured more by the anticipated financial return, the 'bottom line'. So John D. Rockefeller, Andrew Carnegie, and their ilk were worth supporting because your investment was guaranteed to multiply, rather than because the enterprise was exciting 'for Queen and Country'. Early twentieth-century America had gigantic, monolithic Ford . . . while England had a variety of small engineering concerns like Morris Garages (MG).

The other major reason why societies like Victorian England can pick themselves up by their bootstraps and fly is one we've discussed earlier. They lift themselves out of the old constraints, and into a new set of rules. In *The Science of Discworld* and *The Science of Discworld II* we explained why the space bolas, a kind of enormous Ferris wheel in orbit, is capable of carrying people into space far cheaper than rockets – in fact, requiring less energy than anyone would calculate using Newton's laws of motion and gravity. We took one

---

* In their 1980 book *Autopoiesis and Cognition*, Humberto Maturana and Francisco Varela confused this kind of recursion with a life force, and called it 'autopoiesis'. Many self-consciously modern management experts cite this concept, without having the foggiest idea what it is.

further step, and invoked the space elevator, a very strong cable hung from geostationary orbit, which would be harder to build but would require even less energy. The trick is that people and goods coming down can help to lift other people and goods up. The energetics satisfy all the standard mathematical rules, but the context supplies an unexpected source of energy.

These gadgets work better than rockets, but not because these use relativity or other clever new physics like quantum. Or because they don't obey Newton's laws, because they do, to the extent that these are still relevant. Instead, the bolas and the elevator have new invention immortalised into them, so that a spaceman who gets into the cabin of a bolas in thin upper atmosphere from a jet aircraft can shortly afterwards get out of the cabin 400 miles up. Going at the right speed, it so happens, to catch the passing cabin of a 400-mile space bolas, which can deposit him, days later, in the right orbit to catch the 15,000-mile bolas, which deposits him in geostationary orbit, 22,000 miles up, after a couple of weeks. Such machines can be powered by using them to drop valuable asteroid material down to Earth, or (in the case of the bolas) by 'pumping' them like a garden swing, using motors in the middle powered by sunlight and reeling in or letting out the cabin tethers as the bolas rotates.

Once we've made the huge initial investment required to build such machinery, rocket technology becomes largely obsolete, just as animal traction was dispossessed by the internal combustion engine. Sure, you can't attach 500 horses to the front of a big canal-barge, because there wouldn't be room on the towpath – but a 500-horsepower marine engine is another matter entirely. Sure, a rocket would use far too much fuel to be a practical method for hoisting goods and people into orbit *en masse* – but that's not the only way to get them there. Yes, Newton's laws still have to be obeyed, and you have to 'pay' to set everything up, and it still costs just the same energy to get people into orbit. But nobody pays once the machinery is there. If you don't believe this, go up in an elevator in a skyscraper, noting how the counterbalance weights go down, and return to solid ground. *Then*, to ram the message home, walk up the stairs.

The wordprocessor we're using to type this book is a metaphorical space elevator compared to a manual typewriter (remember those? Maybe not). A modern automobile is a space elevator compared to a Ford model T or an Austin-7, which were themselves bolases, while 1880s steam cars were rockets. Think of the investment that went into the Victorian railway system, the canals – then realise how this immense investment changed the rules, so that later generations could do all kinds of things that were impossible to their forebears.

Victoriana, then, was not a situation, it was a process. A recursive process, which built itself new rules and new abilities, as previous hard work and innovation led to new capital, new money, and new investment. The new poor, downtrodden though they may have been, were much better off than the rural poor had been. Which is why people poured into the cities where their lives, even though Dickensian, were easier and more interesting than they had been in the countryside. The urban newcomers provided a new workforce to build new industries. They provided a useful consumer base too. Those workmen's cottages, still found in the suburbs of many towns, were not only housing for an exploited labour force; they were also a source of new wealth for that young aristocrat back from the Gold Coast who'd opened a pickle factory in Manchester. He had seen the sauces made in Madagascar or Goa, liked the taste, and thought that he could sell them to workmen to put on their sausages and bacon. Think of him for a moment, perhaps a chinless wonder who employed thirty men to mix the tropical-fruit ingredients and boil them in great cast-iron vats. The vats had been made in Sheffield and

---

* When Jack started at Birmingham University in the 1950s, the factory behind the university made cooking pots for missionaries, just like the ones that were popular in the cartoons in *Punch* – with the missionaries as ingredients, not cooks, you appreciate. No doubt the same factory had once made the original sauce vats used in Madagascar and Goa.

carried by narrow-boat along canals, giving coin to perhaps fifty workmen who supplied the original vats and buildings:* His pickle company supported a whole small industry for generations: supplying coke for heating, imported and locally grown fruit and spices to be processed into sauces, special water, glass bottles, printed labels . . .

There would have been half a dozen middle-aged matrons busy at different tasks in his factory, too, even bossing some of the men. This was new – outside the home, anyway. Women also got jobs with him as cleaners, perhaps as secretaries to some of the senior staff, and women earning their own money was a massive wedge driven into a male-dominated society. In that society, it was rare even for courtesans to have control of their own funds, to that extent Mimi in *La Bohème* is more realistic than Flora in *La Traviata*. The laws and customs then were very different from what we accept as 'normal' now: young women and older ones were exploited sexually, large numbers of workmen died from industrial accidents and pollution.* Only through their suffering – and their triumphs – could the next generation be built.

Today's Britons are an integral part of this onward and upward process, and in order to see why the triumphs of our real Victorian history have lessons for us now, we must understand what happened then.

There was one major difference, among millions of individual tiny differences, between Victorian Britain and Russia (or China). The

---

* Details can be found in many personal diaries, such as those kept by the foremen of the spinning and weaving mills in Lancashire as exercises in writing for their evening classes. We learn that sexual engagement with women employees was sometimes necessary for these men, in order to retain the respect of their colleagues, to maintain obedience by the workforce, even when they found it horrible themselves. In the armed forces, of course, and in prisons, the social 'rules', the peer pressure to sin grossly, were too powerful to resist, too awful for us to contemplate now.

British had several sources of social heterogeneity, dissidence, of exposure to the public eye of things being done or understood in different ways. From the Baptist chapel to the Quaker meeting house, from the Catholic cathedral with its sweet music and incomprehensible prayers to the Jewish synagogues with their strangely cloaked and hatted congregants who turned into your lawyer or your accountant during the week, religion was obviously diverse. In Poland and Russia, there were pogroms (particularly during the late nineteenth century); in England, there were only taxes. Even in English prisons, very different religious practices were respected, perhaps as much in the breach as in the practice, but the theory was well known and encouraged – if not enforced – by the law. This freedom of thought, word and deed lasted. After the Second World War, after the defeat of Nazism at immense cost, with London still in ruins and food rationed, Sir Oswald Mosley was an avowed fascist whose Blackshirts came down to the East End of London to promote their racist views. Jack was involved in street fights with them about once a month. Even then, he was pleased that their horrible speeches were permitted by the law. In the USA or Russia, Mosley would either have been in jail or elected president. There was a context of heterogeneity, of difference being more than accepted, being valued with a smile. And this was part of an unbroken tradition, going back to Victorian times.

The big difference that made Victorian Britain successful, itself fostered recursively by all the success stories within it – and by the disparate nature of these successes, such as Quakers, railways, big beautiful bridges, fewer starving children, control of some diseases – was in the ambience, the context, which *promoted difference*. It has been fashionable for a particularly naive kind of historian of science to point to the social context of scientific theories, and to pretend that science is therefore entirely socially driven. It is usually claimed, by the same token, that this provenance denies science its authority, so its truths merely follow social convention.

Victorian evolutionists provide a precise refutation of that view.

Wallace, for example, was born to poor parents, was apprenticed to a watchmaker for a while (obviously one of our wizards had been instructed to achieve this), then became a successful – though indigent land agent, then a more successful animal and plant collector. He never made enough money to join the upper middle class, even after his star had risen alongside Darwin's.

Darwin was a junior aristocrat, his parents were well off, and it would have been entirely proper for him to have become a curate – and, indeed to have written *Theology of Species*. Other pro-evolutionists, as various as Owen (mistaken by Darwin for an anti-evolutionist because of his careful analysis of the anatomical implications of the Darwin/Wallace natural-selection idea), Huxley, Spencer, Kingsley, were all from different strata of society. We have seen that the first printing of *Origin of Species* was inadequate for the market, and all copies were sold by the second morning after publication. Would that have happened in nineteenth-century India? In Russia under the czars, *or* after the revolution? In the United States . . . possibly. And in the German part of Prussia. Dickens's stories, critical as they were of the existing order, were anxiously awaited by all strata of society in England – and by many in the eastern United States.

It would not have been quite so strange if this heterogeneous society involved different groups that picked up on *different* ideas, according to their various philosophies and theologies. However, what really happened, both to Dickens and to Darwin and later to Wells, was a very general appreciation of their radical ideas, very widely, across all of those diverse groups. The same alternative views were welcomed by many different strata of society. More so, perhaps, than in any other society since, heterodoxy was almost the rule. Working men's clubs were hotbeds of rational argument, thanks to the establishment of evening classes by the Workers' Educational Association. Education for the common man was promoted by the new technical colleges and the British Association for the Advancement of Science.

To some extent, the same went for all the embryo universities which, in Victorian times, had been seeded by philanthropic discussion groups in the big cities. These establishments, dark red-brick buildings found in the centres of all English industrial cities, were very different organisations from the ancient universities. The other half of the building, or the building opposite on the same street, was often the public library, an organisation not to be found in Russia or China at that time. These organisations provided a way up from manual labour to artisan, and there were a thousand such establishments all over Victoriana.

The real universities, of Oxford, Cambridge, Edinburgh, St Andrew's, were promoting orthodoxy via classics and the literary and governmental arts. The sciences were slowly coming in, mainly as theoretical physics and astrophysics, which needed only brains and blackboards, like mathematics. Practical sciences like geology and palaeontology, chemistry, and zoology went on in dark and dirty laboratories with tall glass and dark wood partitions; botany was backed up by aromatic herbaria. Such work had a very low status compared to mathematics and philosophy – it had associations with manual labour and dirt. However, archaeology, because of its continuing association with the classical world and its artefacts, had quite high status.

The burgeoning middle class didn't, by and large, aspire to these arcane practices. They wanted technical and scientific information, not to potter about with theories, however important and romantic. They didn't want classical anything, certainly not the classics. The universities proper were still requiring a classical education of all aspiring students, and even in the 1970s they continued to require competence in a foreign language from science entrants (as evidence, presumably, of *some* culture – they never required science or mathematics from arts or classics entrants). The workmen and the artisans' guilds cooperated to produce the apprenticeship system, and this was in many ways the model for their own educational organisations.

These, notably the WEA, provided exactly what was wanted, guided and monitored by the artisans' guilds and by the elected council representatives who helped oversee their relations with local industry, especially apprenticeship schemes. 'City and Guilds' examinations, granting certificates and diplomas, were the educational currency of these self-organised educational systems, and they continued until the 1960s. They were the labels that qualified erstwhile labourers as artisans, worthy of respect by their peers.

This pulling yourself up by your bootstraps into respectable citizenship contrasts with the attitude to elected local councilmen by the universities that these organisations matured into. Like the ancient universities, new ones like Birmingham and Manchester rewarded local elected dignitaries, mayors, and councillors with honorary degrees. These empty titles, contrasting both with the earned certificates of the artisans and with the honorary degrees given to eminent scholars in recognition and respect, ensured a political allegiance – and devalued academia in general. Unfortunately, the profusion of such young universities in late twentieth-century England has meant that non-technical, even non-scientific subjects have again become fashionable, to the exclusion of that artisan education which was so healthy in late Victorian times. The devaluation of academic degrees of all kinds has continued apace, but at the same time the alternative and more worthy routes to self-advancement have atrophied.

Does this matter?

Indeed it does. Perhaps Owen Harry, who had himself risen from a poor Welsh beginning near Cardiff's Tiger Bay to become a very young chief technician in Jack's zoology dept at Birmingham University, and later became a senior lecturer at Belfast University, put this best when he described its main negative consequence as 'a lack of sergeants'.

There is a story about officer training and examination in the British Army in the 1950s. One of the most important questions was 'How do you dig a trench?'. The correct answer was 'I say "Sergeant, dig me a trench!"' Sergeants are people who organise the *doing*. They are not experts in what to do, or when: that's the prerogative of officers, who theoretically constitute the brains of the organisation. Officers decide what has to be done, but don't know how to do it. Sergeants don't actually *do* things, either, except occasionally when they have to. Their role is to organise squads of ignorant men, often incompetent, but well trained to obey orders, so that they cooperate effectively. Sergeants are the layer that *makes* cooperation effective: they know how to get things done. Privates know how to do what they're told, and are trained not to do anything else.

We didn't say efficient; it's a common mistake to see efficiency as something to be striven for. Efficiency is a concept borrowed from engineering and physics, a measure of how much you get out for how much you put in. Sergeants are in some respects the least efficient way of getting things done; they have a tendency towards repetition and sarcasm, confident that a few of their recruits will graduate from basic training with some degree of competence. But sergeants are *very* effective, and the system they are part of is very robust.

Darwin and Wallace, Spencer and Wells, all came up through a system that was very robust in this way. All of them, different as they were, knew that writing books was a prime way of affecting the society around you. There was no television, no films, and only a fraction of people went to the theatre or the opera . . . mostly to music hall and pantomimes around Christmas. Dickens, Kingsley, the Brontë sisters, and Thomas Hardy made people – lots of people – think new thoughts and lead new lives. The working men's clubs and their links with the public libraries brought reading skills to a higher level than ever before.

So this audience was ripe for persuasive texts that could take them out of simple biblical knowledge into new theologies, even into atheism. Huxley, 'Darwin's Bulldog', promoted Darwinism as the antithesis of a God-made world. From the aspiring middle class of Victoriana grew our modern secular age, with God relegated to the plaything of a few of the less modern clergy. Modern clergy don't believe in a twelve-foot Englishman up there in the sky, with Heaven as an eternal Buckingham Palace garden party. Particularly from those French philosophers who continued sophisticated theological criticism in lineages derived from Voltaire, our clerics learned to do without that strong Victorian style of Christianity. That form of Anglicanism, confident that God really was looking after the English, didn't need to embarrass itself with overt prayers. The rituals would suffice (provided they weren't noisy like the Welsh, or showy like the Catholics).

We have lost strong simple religion, we have lost academic excellence, we have gained a secular society that maintains the heterogeneity that made it so robust in Victorian times and later. However, we are now pursuing policies, particularly in education, that fail to provide society with all those able people who built the Victorian and Edwardian edifices, both material and theoretical.

There are routes away from this pessimism. In *The Science of Discworld II* we referred to humans as *Pan narrans*, the storytelling chimpanzee. Our overall message was that humans need to make stories to motivate themselves, to identify goals, and to distinguish good from evil.

Here we go a step further.

Technological and Civilised Man, we believe, must become *Polypan multinarrans,** to extend the metaphor rather further. Human beings must become ever more diverse, valuing and enjoying each other's differences rather than fearing them or suppressing them. And mere explanation is not enough. To gain understanding, a useful working philosophy as appropriate for action as for judgement and

---

* Sorry, it's one of those horrible Graeco-Latin hybrids. But, like 'television', it's comprehensible.

decision, an explanation is only rarely good enough. People find simple explanations satisfying because they enable thin causal chains of the kind we build for our own personal memories and causalities. But the real world, even the world of other people and their likes, dislikes, and prejudices – sometimes so rigidly held that our own lives and those of our loved ones don't matter to them – doesn't work like that.

We owe it to ourselves, and to those for whom we are responsible and those who respect us, to develop multi-causal understanding. We can do that, as suggested here, by simultaneously encompassing *several* explanations of each puzzle, explanations that disagree productively with each other. *Multinarrans*: many stories. So one person, even a Newton or a Shakespeare or a Darwin, will not really be enough, despite the story we have just told you. Our fictional Darwin is a symbol for an endless stream of Darwins, challenging orthodoxy *and being right*, a glorious network of innovative thinkers and radicals. People who try to keep ancient cultures alive by blowing up the competition achieve nothing, except widespread contempt for their objectives. They doom their own enterprise by their methods, and they betray a terrible lack of confidence that what matters to them can survive without coercion and violence.

Back to sergeants, and the way things are really done: 'Sergeant, dig a trench.' This is how *Polypan multinarrans* gets things done. How many people are needed to understand a jet airliner? To build one? Recursion in technology really is like biological evolution, it really does expand the phase space. It expands it so much that most of us have virtually no understanding of how the world we live in works. In fact, it is essential that we don't, because there would be too much for anyone to understand.

But we do need to understand that this is what the world is like. Otherwise we don't just lose the sergeants: we lose the ability to build aircraft that fly, dishwashers that clean, cars that don't pollute (as much). We stop being able to cure (some of) the sick, to feed (most of) the planet, and to house, clothe, and wash a burgeoning humanity.

Our world is changing, and it's changing very fast, and we ourselves are the inescapable agents of that change. If we stagnate, like our fictional Victoriana, we die. Staying where we are is not an option. Static resources cannot continue to support us.

We make our world work by introducing new, undreamt-of rules and possibilities, by considering alternatives and making decisions, which feel like 'free will', and work that way, even if they are 'really' deterministic. We build on the present to create a bigger future. Science standing on technology, and technology standing on science, provide a successful ladder that leads to extelligence.

Is it, perhaps, the *only* one?

The past was another country, but the future is an alien world.

And yet . . .

The most remarkable thing about the universe, as Einstein once said, is that it is comprehensible. Not in every aspect, but in enough to make us feel at home in it. It *makes sense* – almost as much as a Discworld story. Which is amazing because facts don't have to make sense: only well-crafted fiction has to obey such rigid rules.

Part of this comprehensibility can be explained. We evolved in the universe, and we evolved to survive in it. Being able to tell ourselves 'what if' stories about it – to understand it – has survival value. We have been selected, by nature, to tell such stories.

What is less easy to explain is why the universe can be represented by human stories at all. But then, if it wasn't, we wouldn't be telling them, would we?

Which brings us back to Charles Darwin, architect of our own present, which was his future, and would surely seem alien to any Victorian. In Chapter 18 we left him sitting on an 'entangled bank', watching birds and insects, and musing on the nature of life. The final paragraph of *The Origin*, which began with gentle musings about entangled banks, now works its way to its revolutionary conclusion:

> From the war of nature, from famine and death, the most exalted object which we are capable of conceiving, namely, the production of the higher animals, directly follows. There is grandeur in this view of life, with its several powers, having been originally breathed into a few forms or into one; and that, whilst this planet has gone cycling on according to the fixed law of gravity, from so simple a beginning endless forms most beautiful and most wonderful have been, and are being, evolved.

TWENTY-FIVE

THE ENTANGLED BANK

 IT WAS MIDNIGHT IN THE museum's Central Hall when the wizards appeared. There were a few lights on; just enough to see the skeletons.

'Is this a temple of some kind?' said the Chair of Indefinite Studies, patting his pockets for his tobacco pouch and a packet of Wizlas. 'One of the weirder ones, perhaps?'

+++ Indeed +++ boomed the voice of Hex from the middle air. +++ In all the universes of the *The Ology*, it was the Temple of the Ascent of Man. Here, it is not +++

'Very impressive,' muttered the Dean. 'But why don't we just show him the big snowball? He'd be pretty pleased to know it was because of him humans got away.'

'We've scared the poor chap enough, that's why!' snapped Ridcully. 'He'll understand *this*. Hex says they started building when Darwin was alive. Stuffed animals, bones . . . it's the kind of thing he knows. Now stand back and give the chap some air, will you?'

They stepped away from the chair on which Charles Darwin had been transported, wreathed in the blue light. Ridcully snapped his fingers.

Darwin opened his eyes, and groaned.

'It never ends!'

'No, we're sending you back, sir,' said Ridcully. 'That is, you'll soon wake up. But we thought there is something you should see first.'

327

'I've seen enough!'

'Not quite enough. Lights, gentlemen, please,' said Ridcully, straightening up.

Light is the easiest magic to do. A glow rose in the hall.

'The Museum of Natural History, Mr Darwin,' said Ridcully, standing back. 'It opened after your death at a venerable age. It's your future. I believe there is a statue to you here somewhere. Place of honour, no doubt. Please listen. I would like you to know that because of you, humanity turned out to be fit enough to survive.'

Darwin stared around at the hall, and then looked askance at the wizards.

'The phrase "survival of the fittest" was not—' he began.

'Survival of the luckiest in this case, I fear,' said Ridcully. 'You are familiar with the idea of natural catastrophes throughout history, Mr Darwin?'

'Indeed! One only has to examine—'

'But you will not have known that they wiped intelligent life from the face of the globe,' said Ridcully, sombrely. 'Sit down again, sir . . .'

They told him about the crab-like civilisation, and the octopus-like civilisation and the lizard-like civilisation. They told him about the snowball.*

Darwin, Ponder thought, bore up well. He didn't scream or try to run away. What he did do was, in a way, worse: he asked questions, in a slow, solemn voice, and then asked more questions.

Strangely, he kept away from ones like 'how do you know this?' and 'how can you be so sure?'. He looked like a man anxious to avoid certain answers.

For his part, Mustrum Ridcully very nearly told the whole truth on several occasions.

At last Darwin said, 'I think I see,' in a tone of finality.

'I'm sorry we had to—' Ridcully began, but Darwin held up a hand.

---

* See *The Science of Discworld* and *The Science of Discworld II*.

'I do know the truth of all this,' he said.

'You *do?*' said Ridcully. 'Really?'

'Indeed, a few years ago there was a rather popular novel published. *A Christmas Carol*. Did you read it?'

Ponder looked down at the hitherto blank piece of paper on his clipboard. Hex had been told to be quiet; Charles Darwin was probably not in the right frame of mind for booming voices from the sky. But Hex was resourceful.

'By Charles Dickens?' said Ponder, trying not to look as though he was reading the writing that had suddenly filled the page. 'The story of the redemption of a misanthrope via ghostly intervention?'

'Quite so,' said Darwin, still speaking in the careful, wooden voice. 'It is clear to me that something similar is happening to me. You are not ghosts, of course, but aspects of my own mind. I was resting on a bank near my home. I had been wrestling at length with some of the perturbing implications of my work. It was a warm day. I fell asleep, and you, and that . . . god . . . and all this, are a kind of . . . pantomime in the theatre of my brain as my thinking resolves itself.'

The wizards looked at one another. The Dean shrugged.

Ridcully grinned. 'Hold on to that thought, sir.'

'And I feel sure that when I awake I *will* have reached a resolve,' said Darwin, a man firmly nailing his thoughts in order. 'And, I fervently trust I will have forgotten the means by which I did so. I certainly would not wish to recall the wheeled elephant. Or the poor crabs. And as for the dirigible whale . . .'

'You *want* to forget?' said Ridcully.

'Oh, yes!'

'Since that is your clear request, I have no doubt it will be the case,' said Ridcully, glancing questioningly at Ponder. Ponder glanced at the clipboard and nodded. It was a direct request, after all. Ridcully was, Ponder noted, quite clever under all that shouting.

Apparently relieved at this, Darwin looked around the hall again.

'"I dreamt I dwelt in marble halls", indeed,' he said.

The words 'Reference to a popular song written by Michael W. Balfe, manager of the Lyceum Theatre, London, in 1841' floated across Ponder's clipboard.

'I don't recognise some of these very impressive skeletons,' Darwin went on. 'But that is Robert Owen's *Diplodocus carnegii*, clearly . . .'

He turned sharply.

'Humanity survives, you say?' he said. 'It rode out to the stars on tamed comets?'

'Something like that, Mr Darwin,' said Ridcully.

'And it flourishes?'

'We don't know. But it survives better that it would under a mile of ice, I suspect.'

'It has a *chance* to survive,' said Darwin.

'Exactly.'

'Even so . . . to trust your future to some frail craft speeding through the unknown void, prey to unthinkable dangers . . .'

'That was what the dinosaurs did,' said Ridcully. 'And the crabs. And all the rest of them.'

'I beg your pardon?'

'I meant that this world is a pretty frail craft, if you take the long view.'

'Ha. Nevertheless, some vestige of life surely survives every catastrophe,' said Darwin, as if following a train of thought. 'Deep under the sea, perhaps. In seeds and spores . . .'

'And is that how it should be?' said Ridcully. 'New thinkin' creatures arisin' and being forever smashed down? If evolution didn't stop at the edge of the sea, why should it stop at the edge of the air? The beach was once an unknown void. Surely the evidence that mankind has risen thus far may give him hope for a still higher destiny in the distant future?'

Ponder looked down at his clipboard. Hex had written: he is *quoting* Darwin.

'An interesting thought, sir,' said Darwin, and managed a smile. 'And now, I think, I really should like to awaken.'

Ridcully snapped his fingers.

'We *can* get rid of those memories, can't we?' he said, as the blue glow enveloped Darwin yet again.

'Oh yes,' said Ponder. 'He's asked us to, so it's ethically correct. Well done, sir. Hex can see to it.'

'Well then,' said Ridcully, rubbing his hands. 'Send him back, Hex. With perhaps just a tiny recollection. A souvenir, as it were.'

Darwin vanished. 'Job done, gentlemen,' said the Archchancellor. 'All that remains now is to get back for—'

'We ought to make sure there are no more Auditors left on Roundworld, sir,' said Ponder.

'On that subject—' Rincewind began, but Ridcully waved him away. 'That at least can wait,' he said. 'We've established the time-line, it's nice and stable, and we can—'

'Er, I don't think they *want* to wait, sir,' said Rincewind, backing away.

Shadows were pouring in to the Central Hall. Over the double stair-case, a cloud was forming. It looked like the grey robe of an Auditor, but hugely bigger, and as the wizards watched the greyness dark-ened to coal-mine black.

The bloated shape drifted forward, while hundreds more of the empty grey robes continue to merge with it.

'And I think they're a bit angry,' Rincewind added.

Trailing greyness after it, filling the hall from edge to edge, the Auditor bore down on the wizards.

'Hex—' Ponder began.

'Too late,' boomed the Auditor. '*We* have control now. No magic, no science, no chocolate. We have to thank you for this place. Never was there a species so determined to destroy itself. In this world, we can win without trying! Do you know the wars you've unleashed upon this toy world? The plagues, the famine, the whole *science* of death? Are you not ashamed?'

'What's he talking about, Stibbons?' said Ridcully, not taking his eyes off the cloud.

'There are a number of wars in the next couple of hundred years, sir,' said Ponder. 'Big ones.'

'Darwin's fault?'

'Er, sir.'

'Just "er", Stibbons?'

'"Er" is a very precise term in this context, sir. It means we don't have time for a big debate. But certainly the wars are bigger and more frequent than the ones that took place in the world of *The Ology*.'

'Bad thing, then?' said Ridcully, who liked his philosophy succinct.

'Er again, sir, I'm afraid,' said Ponder.

'Care to expand?'

'In short, sir, more people will die in wars, far fewer will die of disease and medical problems of all kinds. And humanity survives the snowball. The first humans left the planet in converted weapons of war, sir.'

'That's monkeys for you, Stibbons,' said Ridcully. He looked up at the cloud of pure Auditor.

'No, we're not ashamed,' he said. 'Humans get a chance to go on.'

'They won't have earned it!'

'Strange that this concerns you,' said Ridcully.

'Do you know the terrors that will confront them?' the Auditor demanded. 'And the terrors that they will bring *with* them?'

'No, but I doubt if they're worse that the ones they've met already,' said Ridcully. 'Anyway, you don't *care* about them. You just want them to die quietly. Don't you?'

The Auditor shimmered. Ponder wondered how many Auditors had come together to create it. It seemed, now, to be hesitant, unsure . . .

It said: 'I want . . . I . . . '

. . . and exploded into fog which, itself, faded away.

'Not learned quite enough, then,' said Ridcully, and sniffed. 'Well, let's send Darwin back and go home, shall we? I'm sure we've missed at least one meal. Where's Rincewind?'

+++ Hiding in the Minerals Gallery +++ said Hex.

'Impressive. I didn't even see him move. Oh well, I dare say you can pick him up later. Let's go.'

'What did it mean by the terrors they bring with them?' said the Dean.

'Well, they're still monkeys,' said Ridcully. 'Still screaming at one another, trailing all that evolution behind them, wherever they go.'

'Darwin said something like that, sir. In *The Descent of Man*,' said Ponder.

'Good chap, Darwin,' said Ridcully. 'Would have made a good wizard.'

'Did you know they put his statue in the canteen, sir?' said Ponder, a little shocked.

'Did they? Good idea,' said Ridcully brusquely. 'That way, *every* sensible person *sees* it. Ready, Hex.'

And the Central Hall was empty again, apart from the fossils.

Charles Darwin awoke. For a moment so brief that a blink ended it, there was a sense of complete disorientation. But then he sat up, feeling unaccountably exhilarated, and looked around at the tangled, busy bank, with its birds and flitting insects, and thought: Yes. That's right. That's how it is.

AFTERTHOUGHT

The Darwin family motto:
cave et aude.
Watch, and listen.

INDEX